JET NOISE IS THE SOUND OF FREEDOM

THINGS AIRMEN KNOW

BY JIM PFAFF

JET NOISE IS THE SOUND OF FREEDOM
Things Airmen Know

© 2025 Jim Pfaff

All rights reserved. No portion of this book may be reproduced in any fashion, print, facsimile, digital or electronic, or by any method yet to be developed, without express permission by the copyright holder.

ISBN 979-8-9931698-3-5

For information, contact the author:

Jim Pfaff
jp135nav@gmail.com

Published by:

Chilidog Press LLC
pbronson@chilidogpress.com

Chilidog Press LLC
Loveland, Ohio
www.chilidogpress.com

Cover Design and Book Layout by Andy Melchers

Unless otherwise credited, all photos are from the United States Air Force, via Wikimedia Commons.

JIM PFAFF

TABLE OF CONTENTS

BEGINNINGS	13
WORLD WAR I	19
BETWEEN THE WARS	23
WORLD WAR II	33
JET AGE: COLD WAR AND KOREAN WAR	69
VIETNAM	87
COLD WAR AND A DANGEROUS WORLD –1970s TO 1991	105
BOOKS ABOUT THE AIR FORCE	109
CELEBRITIES IN AIR FORCE BLUE	113
MOVIES ABOUT THE AIR FORCE	121
FICTIONAL AIRMEN	127
RANKS, INSIGNIA, UNIFORMS	131
AIR FORCE CULTURE AND LORE	141
THINGS WE SAY: "COMMON KNOWLEDGE"	163
BRATS	185
MISSION AND WORK	187
FIRSTS AND LASTS, SUPERLATIVES	213
A CHANGING WORLD	221
WAR ON TERROR	233
COME FLY WITH ME	239
TOWARD THE FUTURE	251
ACKNOWLEDGEMENTS / ABOUT THE AUTHOR	255
PHOTO CREDITS	257

JIM PFAFF

DEDICATION

*For my family, especially Karin and Johanna,
who saw me fly away so many times.*

*May your children live in a peaceful
and prosperous world.*

JIM PFAFF

PREFACE

This book is an overview, highlighting the history, lore and culture that shape the Air Force. You might be aware of much of this. If you know someone who wears Air Force blue, or who wants to, this may help you to understand them better. If you have a blue uniform in the back of a closet, you may find a small piece of your story here, from the perspective of one who has been a part of the Air Force for a long time.

When I was seven years old, my Cub Scout pack visited the Air Force Museum. "Ancient" relics of the World Wars astonished me; modern aircraft intrigued me. Soon after, my family went to see my Uncle Bill's family in Indiana. He was a technical sergeant in the Air Force, and they had relocated from England. We visited his base and he took us to a building near the end of the runway to watch a four-ship formation launch of B-58 bombers. The windows rattled as the powerful engines drove the jets into the sky, all four in less than a minute. They roared into the blue, and I was hooked! At the base exchange, my uncle bought me a toy B-58. I still have it.

I was an awkward and clumsy boy, nowhere near the sleek athletic type the Air Force Academy seeks. However, I was awarded an Air Force ROTC scholarship. The dream of flight was still there, but my eyes weren't quite good enough to be a pilot. "Line 9" on the eye test indicated 20/20. I can still hear the technician asking me to read that line twice, then saying, "Let's try Line 8." I could read that slightly larger font, so I qualified to be a navigator.

A toy B-58 that started a journey.

Upon graduation from navigator training in California, I was assigned to a Strategic Air Command (SAC) KC-135, the air refueling tanker. It was the least glamorous job in the least glamorous airplane in the Air Force, but it was mine. My first flying assignment was at the same base where we visited my uncle in Indiana, just a few hours from home. I qualified in our unit's missions, became an instructor navigator, then transferred to Alaska, where I flew the RC-135, a reconnaissance aircraft. In the Cold War, this was an exciting mission. However, the Soviet Union's collapse in 1991 meant the

United States no longer had to stand ready for nuclear war at a moment's notice. The Global Positioning System (GPS) was coming into use then, too, and the Air Force no longer needed navigators with sextants, charts, plotters, and dividers to guide aircraft across oceans and the Arctic.

Significant reductions were coming to the Air Force, particularly in my career field. Expecting the changes to be implemented in the active force first, I joined the Ohio Air National Guard (ANG) KC-135 unit based in Columbus, Ohio, looking for civilian employment and hoping to extend my flying career. ANG units have a handful of full-time slots, and I was selected for this role. The "we're a family" aspect of the ANG unit suited me. The active Air Force is dynamic, with a constant flow of people coming and going. Outgoing personalities and athletic types thrive. I tend to be shy and accomplish my work quietly.

Still, change was coming, and when ANG navigator jobs were being eliminated, I was offered a position at Air Mobility Command (AMC) headquarters at Scott Air Force Base (AFB), Illinois. We loved the St. Louis area and thought I might retire in this role. But my wife had a rare blood cancer, and the best treatment location was at the renowned James Cancer Hospital in Columbus. I was able to transfer back to the Ohio ANG tanker unit as an aircraft maintenance officer. I was promoted to colonel and eventually retired. My military career led to teaching Air Force Junior Reserve Officer Training Corps (AFJROTC) for twelve years.

So, this book is a look at the world I've inhabited. It's not a complete history, but rather what I would mention if we toured a museum or a base together. Our history drives the force forward. Airmen know that the past inspires and teaches us. Our predecessors left us a mission to continue.

Like any branch of the military, the Air Force is a unique society, a profession that is often more a calling than a job. We will visit a base and go out on the flightline. The Airman's language occasionally veers toward dialect, so some explanation is warranted.

Brigadier General Chuck Yeager was a fighter pilot, a war hero, a test pilot, and the first human to exceed the speed of sound. I have little in common with all that. I flew aboard airliner-sized aircraft and my impact was far less. But as Chuck said, "All that I am... I owe to the Air Force."

CHAPTER 1
BEGINNINGS

"The future of our nation is forever bound up by the development of airpower." - Billy Mitchell

The sight of a hot-air balloon on a summer evening is pleasant, a dreamlike vision in the sky. We sit on the grass and stare, our thoughts drifting like the balloon's path. Imagine the wonder these giants inspired in a time when flight was only a fantasy.

In the summer of 1861, as the Civil War began, a scientist named Thaddeus S. C. Lowe approached President Abraham Lincoln, offering his balloon for reconnaissance. Lincoln, always intrigued by technology, appointed him Chief Aeronaut. Lowe flew his hot-air balloon at Bull Run and other battles. Many Army officers saw this as a waste of money, and in 1863, Lowe was relieved of his duties and returned to his scientific pursuits.

Thaddeus Lowe had another connection to aviation history: his granddaughter, Florence, is remembered as Pancho Barnes, an air racer and stunt pilot in the 1920s and 30s. She owned a bar near Muroc Army Air Field (later renamed Edwards Air Force Base) that was immortalized in Tom Wolfe's *The Right Stuff* as the hangout of the test pilots.

The Balloon Corps was revived during the Spanish-American War in 1898. Lt. Col. Joseph Maxfield and Sgt. Ivy Baldwin took the Army's only hot-air balloon to Cuba, providing reconnaissance for the attack at San Juan Hill. Unfortunately, in a balloon, if you can see the enemy, they can see you. The balloon was quickly shot full of holes. The Balloon Corps disbanded in 1899.

However, in pre-World War I

First Airship in US military service

Europe, airships advanced faster than airplanes. (Zeppelins and dirigibles are airships with a rigid framework.) Commercial airship service operated in Europe by 1910, and thousands of passengers had flown by 1914. At that time, only a few hundred intrepid souls had flown in flimsy wood and fabric "aeroplanes" by ones and twos—and many had come to unfortunate ends.

In 1907, military aviation was placed under the Signal Corps, home to some of the top scientific minds of the day. A. W. Greely was the Chief Signal Officer (CSO) from 1887 to 1907. Wounded at the Battle of Antietam in the Civil War, he led pioneering Arctic expeditions and installed telegraph lines from Alaska and the Philippines to the United States. He supported airship development. Greely was awarded the Medal of Honor for his lifetime of accomplishments, an almost unheard-of distinction.

James Allen succeeded Greely. He brought wireless telegraphy and radio to the Army. He funded the first Army airplanes. George P. Scriven became CSO in 1913. He became chairman of the National Advisory Committee on Aeronautics (NACA), the forerunner of NASA. He served until 1917.

His successor, George O. Squier, held a PhD from Johns Hopkins and wrote his dissertation on the electrochemical effects of magnetization. He held more than 60 patents, invented a high-speed camera, and developed telephone multiplexing, enabling multiple signals to be carried on the same wire. In September 1908, Squier became the first military passenger to fly in an airplane. He retired in 1923 and invented Muzak, a background music provider that became ubiquitous in American business and retail establishments throughout the 20th century.

In August 1908, the Army purchased its first aircraft, Signal Corps Airship No. 1, a dirigible designed and built by Thomas Baldwin, a circus acrobat and balloonist. His airship could make a two-hour flight at 20 miles per hour. It required two crewmen, a pilot and a rudder operator. They rode on a gondola beneath the balloon, with a 500 pound cargo or passenger payload capacity. The Army's pilots were Lts. Benjamin Foulois and Frank Lahm, and Capt. Charles Chandler.

Baldwin designed airplanes with Glenn Curtiss and trained airplane pilots, including Billy Mitchell. When the United States entered World War I, Baldwin was commissioned a captain and inspected lighter-than-air craft procured by the Army during the war.

Even with the airplane's success and fame in the war, the military held onto using lighter-than-air craft for years. Airships had some promise: they offered superior range and passenger capacity well into the 1930s. However, accidents like those that befell Navy airships between 1924 and 1933, and the crash of the Hindenburg in 1937, diminished confidence in airships, while airplane technology advanced. The Navy used blimps into the 1960s for search and observation missions.

THE WRIGHT BROTHERS

Airplane aviation begins with the Wright brothers, who first flew on December 17, 1903, at a barren stretch of the Outer Banks called Kitty Hawk, North Carolina. Orville and Wilbur Wright had spent four years testing flying contraptions, starting with kites and gliders, looking for a way for humans to fly. Most researchers had been concerned with powering the aircraft; the Wrights, with their background in bicycles, focused more on controlling the craft, which was key to their success. They were just two guys who owned a bicycle shop in Dayton, Ohio. They became interested in flight in the 1890s, especially after the much-publicized crash of glider pioneer Otto Lilienthal in 1896. The Wrights were quietly single-minded about this. As they experimented with kites, gliders, and then powered aircraft, they took their experiments to the Outer Banks for the windy skies, endless sand to land on, and privacy. After all, the prevailing wisdom was that anyone who said men could fly must be a nut!

Orville made the first flight on that December day, a 12-second, 120-foot hop that was the first powered airplane flight. Wilbur got the 59-second, 852-foot flight that capped off the day.

Neither of the Wright brothers ever donned a uniform. Their connection with the Air Force is that they invented the airplane. What would flight be without them? Would the United States have been the dominant nation in aviation?

The First Flight, 1903

The Wrights sold the US government its first airplane in September 1908. For that aircraft and instructional services (training pilots), the Army paid them $25,000. Their plane met the requirements of 40 miles per hour, carrying two people, with a 125-mile range, and was able to land without damage. The airplane was delivered on August 2, 1909, and the Wrights built many of the Army's first airplanes. Today, Wright-Patterson AFB, near Dayton, bears their name. Who was Patterson? Lt. Frank Stuart Patterson, killed in a crash in 1918, was a member of the prominent Patterson family in Dayton. Lt. Patterson's brother founded National Cash Register.

The first military pilot was well, it's not as simple as you might think.

The first Army officers called pilots were balloonists Capt. Charles deForest Chandler, and Lts. Frank P. Lahm, Benjamin D. Foulois and Thomas Selfridge.

When the Signal Corps Aeronautical Division appeared on August 1, 1907, balloon pilot Capt. Charles Chandler was placed in charge of it.

The first Army officer to solo in an airplane was Lieutenant Selfridge in 1908.

The first three pilots trained by the Wright brothers under their contract with the Army were Lieutenants Foulois, Lahm and Frederick Humphreys. The first Army pilot to solo under the Wright brothers' contract was Humphreys.

Benny Foulois is often called the first military aviator, due to his role as a balloon pilot and was the last survivor among these early aviators.[1] Frank Lahm was listed in a 1958 Air Force historical study as the first military pilot.

The first two rated airplane pilots in the US Army were Lt. Thomas Milling and Lt. Henry H. Arnold. Perhaps due to his career success, Arnold was sometimes honored as the first pilot.

The first written orders designating pilots listed eleven men: Capts. Charles de F. Chandler and Paul W. Beck; First Lts. Roy C. Kirtland, Benjamin D. Foulois, Harold Geiger, Samuel H. McLeary, Lewis E. Goodier, Joseph D. Park, Henry H. Arnold; and 2d Lts. Thomas DeWitt Milling and Lewis H. Brereton.

The first two Military Aviator Badges were awarded to Capt. Chandler and Lt. Milling in October 1913; Lt. Arnold also received his that month. The badge was an eagle carrying signal flags, suspended from a bar inscribed, "Military Aviator." Arnold wore his proudly for the rest of his career, even with his silver pilot wings.

What is certain is that all of them embarked on a dangerous mission, and the outcomes were success or death. Park, Geiger, and Selfridge died in

---

[1] In his autobiography, Foulois states that he was the first selected "to go aloft" on the balloon with Baldwin.

airplane accidents.[2] Humphreys left the Army but was promoted to colonel in the New York National Guard; he died in 1941. Chandler made colonel and commanded balloon units in World War I, but he retired due to disability in 1920 and died in 1939. Lahm, Foulois, Milling and Arnold would wear generals' stars.

To say early flight was dangerous is an understatement. Lt. Thomas Selfridge was the first American military aviator to die in a plane crash. He was flying with Orville Wright on September 17, 1908, when a propeller broke, and the split propeller hit a wire bracing the rudder. The wire shattered the propeller, and the rudder came loose, sending the Wright Flier down. It hit hard: Orville had a broken leg and broken ribs, and Selfridge suffered a fractured skull and died a few hours later.

The first enlisted man to die in a plane crash was Cpl. Frank Scott, on September 28, 1912. Scott was a mechanic who volunteered to fly with a lieutenant at College Park, Maryland. The flight was normal for about ten minutes, but the pilot lost control while attempting to land. Scott was killed in the crash. The pilot, Lt. Lewis Rockwell, died a few hours later.[3] Between 1909 and 1914, about one-fourth of the pilots in the Army were killed in aircraft accidents. Only single men were accepted for pilot training, and a 50 percent addition to base pay was offered for those who flew.

Benjamin D. Foulois lived most of the Air Service's early history. He enlisted in the Army in 1898, fought in the Philippines, and was commissioned in 1901. He was the first pilot checked out on the Army's first dirigible in 1908. He was ordered to take the Army's airplane—there was only one in the inventory—to San Antonio in 1910. He corresponded with the Wrights to learn to fly it. This seems like trying to learn to swim by reading about it, but Foulois did it. He also invented the seatbelt (a hair-raising incident prompted that) and added landing gear to replace the skids the Wrights were using. He led the 1st Aero Squadron in Mexico and laid the foundations for the Air Service during the Great War. Foulois made

Benjamin Foulois, pilot

---

[2] Lt. Selfridge was the first of many aviators who died in the line of duty and are remembered with Air Force Bases named for them. Selfridge Air National Guard Base is in Michigan.

[3] Cpl. Scott is the namesake of Scott AFB, Illinois. Dozens of Air Force facilities have borne fallen Airmen's names. Some are no longer active, such as Rockwell Field, California (now part of North Island Naval Air Station).

brigadier general during the war, and while overshadowed by the colorful Billy Mitchell, he was responsible for training and logistics. He became Chief of the Air Service in 1931.

An American Airman did not fly the first air combat mission. Italian Army Capt. Carlo Piazza flew a reconnaissance sortie near Tripoli, Libya on October 23, 1911, against the Ottoman Army during the Italo-Turkish War. The first operational missions flown by Americans in US aircraft would come in Mexico in March 1916.

American Airmen did their first operational flying in what was then called the Punitive Expedition (yes, punitive means "to punish") into Mexico as the mayhem resulting from the Mexican Revolution spilled across the border. Pancho Villa's men killed several Americans in Mexico, then raided Columbus, New Mexico. President Woodrow Wilson ordered Brigadier General John J. Pershing to assemble a division of about 8,000 men to pursue Villa. The US force crossed into Mexico on March 15, 1916, and the first reconnaissance sortie was flown the next day by the 1st Aero Squadron.

Based at Texas City, Texas since March 1913, the Army's first flying unit deployed its eight Curtiss JN-3 "Jenny" aircraft from San Antonio to Columbus, New Mexico by train. They flew into Mexico, but within a few weeks, due to mechanical problems and a lack of spares, they were down to two flyable airplanes. Mountains in northern Mexico ranged up to over 10,000 feet high, and the Jennies couldn't make it that high. The War Department procured some better aircraft, and some of the mission, communications (carrying messages), and observation was accomplished. The Air Service's first operational deployment was none too successful, but lessons were learned. Keeping a flying squadron going at the end of a long supply line in harsh climates and austere conditions would always be challenging. These pioneers did that. Today, the 1st Aero Squadron is the 1st Reconnaissance Squadron at Beale AFB, California, flying the U-2 and the Global Hawk.

CHAPTER 2
WORLD WAR I

"Fighting in the air is not a sport. It is scientific murder."
- Eddie Rickenbacker

Many people saw the "Great War" as a battle for the future of civilization. Some young Americans volunteered for service with the French Air Force's Lafayette Flying Corps. Its most famous squadron, the Lafayette Escadrille, was formed in 1916. The volunteers were predominantly wealthy, Ivy League-educated young men, as a social class largely Francophile. The Americans fought in French uniforms and under French command, scoring the unit's first victory on May 18, 1916. These 200 young Americans were renowned as heroic knights in a noble cause. They downed 199 German aircraft while losing 66 dead. When the United States entered the war in 1917, many of these Airmen transferred into US service, where they would be leaders and instructors among the burgeoning American Expeditionary Force (AEF) Air Service.

American spending on military aviation was minuscule compared to that of the European powers, and that gap widened as Europe became a battlefield. When Woodrow Wilson asked Congress for a declaration of war in April 1917, the Signal Corps Aviation Section had about 1,000 men, only 35 pilots and 55 airplanes. By the Armistice in November 1918, there were 20,000 officers, 10,000 pilots and 175,000 enlisted men. About one-third of this force was in France. There were 740 aircraft in 45 flying squadrons and 23 balloon companies. The US produced airplanes—mostly trainers—but only a few American-built airplanes reached France. American Airmen destroyed 776 enemy planes and 72 balloons. They flew 150 bomb raids, some more than 100 miles behind enemy lines, and dropped about 137 tons of bombs. Billy Mitchell had General Pershing's approval for "strategic" bombing raids in 1919. It is hard to imagine what their impact could have been.

US forces lost 290 aircraft, 47 balloons and 164 Airmen killed in action, with 200 more missing. About 100 men were taken prisoner. Accidents were more lethal: 319 Airmen died in aviation mishaps. Some 335 Airmen died of other causes.

Americans distinguished themselves in the skies above France. Lt. Stephen W. Thompson became the first flier in a US uniform to score an aerial victory. A gunner in a French-piloted airplane, he shot down a German Albatros D.III on February 5, 1918. The first American ace, shooting down five enemy aircraft, was Frederick Libby, who flew with the British. About 100 Americans were aces in World War I; about one-third of those flew with the British or French.

Lt. Frank Luke was a "balloon buster." An Arizona native, he was a wild card, a cocky loner who would take off on his own to fight the Germans. But he was a talented aerial hunter. In Luke's short time at the front, he became a legend. On August 18, 1918, he claimed a German Fokker in his first dogfight. He then went on a spree unequaled by any other American fighter pilot, ringing up 18 victories (14 balloons) in just ten sorties. On the last of these, on September 29, 1918, his SPAD was shot down by ground fire. Luke got off a few pistol rounds at German infantry approaching him on the ground, then died of his wounds. Eddie Rickenbacker declared Luke to be "the most daring aviator and greatest fighter pilot of the entire war." He was posthumously awarded the Medal of Honor.

A different kind of mission was the first to garner two Medals of Honor. On October 6, 1918, Lts. Erwin Bleckley (observer) and Harold Goettler (pilot) flew a reconnaissance mission in support of the "Lost Battalion," a unit that had been isolated by German forces. They flew at low level to locate the trapped doughboys and drop supplies. Despite heavy, close-range fire from German gunners, they made a couple of runs over the isolated Americans, even landing one damaged plane to get another. Hit again, this plane went down. Goettler was killed, but Bleckley, badly wounded, was able to give the Lost Battalion's exact position to French troops before he died.

America's top-scoring ace in World War I, Eddie Rickenbacker, shot down 26 Germans while flying Nieuport 28 and SPAD XIII aircraft in the 94th "Hat-In-The-Ring" Squadron. He was awarded the Medal of Honor and seven Distinguished Service Crosses (DSC). A former mechanic and race car driver who competed in the first Indianapolis 500, Rickenbacker was assigned as a driver for General Pershing and had to lobby Billy Mitchell to become a pilot. He served in engineering and in combat and succeeded in both. After the war, Rickenbacker had a successful business career and survived two major aircraft

Eddie Rickenbacker

mishaps. One was an airliner crash in which he suffered life-threatening injuries; in the other, Rickenbacker survived a ditching in the Pacific Ocean and spent 24 days in a life raft before rescue. He owned the Indianapolis Motor Speedway for several years and ran Eastern Air Lines for decades. Rickenbacker died in 1973 and is buried in a family plot at Greenlawn Cemetery in Columbus, Ohio.

THE BIRTH OF AIRPOWER

In that first air war, three influential theorists envisioned the airplane as a war-winning weapon: Billy Mitchell, Hugh Trenchard and Giulio Douhet. They advocated strategic bombardment, putting aviation on a course to devastate nations. They imagined air weapons striking the sources and infrastructure of military power to end the gruesome nightmare of trench warfare quickly.

Giulio Douhet, an Italian artillery officer, believed that wars were no longer fought between armies but between whole peoples, and future wars would be total and unrestrained, with civilians as legitimate targets. Douhet thought wars could be won by destroying "the enemy's will to resist." The airplane could fly past ground defenses to attack cities, reducing armies and navies to secondary roles. One thing that Douhet espoused is still an airpower dictum: One must achieve command of the air by attacking airfields, parked aircraft and aircraft factories— "destroying the nests and eggs on the ground" rather than having to "hunt flying birds in the air." This meant an offensive campaign was required.

Hugh Trenchard, a British infantry officer turned pilot, shared similar ideas on applying airpower, rooted in his conviction that civilian morale was fragile. However, his impact was to create an independent air force, the Royal Air Force (RAF). Just as sailors lead a navy, he believed airpower must be led, trained and operated by fliers who understand air operations. Trenchard and other Airmen viewed Army leadership as bound to the past, in a world where the pace of operations was that of a horse. Army officers envisioned using flying squadrons like cavalry, with local and piecemeal deployment, negating the power an independent formation could bring to bear on the larger battlespace.

Upon arriving in Europe, Col. William "Billy" Mitchell sought out Trenchard, and the brash American got on famously with the crusty Englishman. The two shared a vision of how aviation could overcome stalemate. They recognized that air superiority was essential, even before the armies closed with one another. They viewed airpower as inherently offensive, with devastating power to destroy the enemy's war effort and to batter his will to fight. These themes would animate Mitchell for the rest of his life.

The Battle of St. Mihiel, in September 1918, was the first operation in the war carried out by an American army under an American commander, and the first attempt at a combined-arms, ground-and-air operation. This idea would shape future warfare. Col. Mitchell led 1,481 planes (about half were American), including 701 pursuit planes, 366 observation planes, 323 day bombers, and 91 night bombers, in the largest air operation of the war. In just five days, Mitchell's planes flew 4,000 sorties, keeping German aircraft at bay while delivering strikes behind enemy lines.

Looking back at the abattoir of the Western Front, aviation had provided a glimpse of heroic honor. The nightmare of the trenches scarred a generation. In many ways, Europe has never recovered from it.

CHAPTER 3
BETWEEN THE WARS

"Not to have an adequate air force now is to compromise the foundations of national freedom and independence."
- Winston Churchill

Billy Mitchell embodied the US Army Air Service in the 1920s. He was born in France while his family vacationed; his father was a senator from Wisconsin. Young Billy joined the Army in 1898 and was commissioned at age 18. Mitchell served well in the Philippines and Alaska before being assigned to Washington, DC, where he witnessed Orville Wright's flights. The Signal Corps controlled Army aviation, but as a major on the general staff, he was deemed too old for Army pilot training at age 36. So, he spent more than $1,400 of his own money to qualify as a pilot. (At the time, a major made $325 a month.)

Billy Mitchell

With the US declaration of war on April 6, 1917, Mitchell was among the first American troops sent to Europe. Fluent in French, he immediately began working with allied aviators and became the first American officer to fly a combat sortie in Europe, tagging along with the French. He was promoted quickly as the AEF expanded, rising to colonel in 1917. In September 1918, he led the air units in the St. Mihiel attack. The success led to Mitchell's promotion to brigadier general on October 14, 1918. He was one of America's best known Airmen of the war.

After that, the impetuous hero expected to run the Air Service. But the command went to non-flying Maj. Gen. Charles Menoher, with Mitchell as deputy. Menoher had commanded the 42nd "Rainbow" Division in the war. His deputy then was Douglas MacArthur. So General Menoher supervised two of the biggest personalities in American military history. It had to be exhausting!

Menoher believed, as did General Pershing, that air forces supported ground forces. Menoher clashed with Mitchell, a man who saw the possibilities of the airplane and who had no small opinion of his own possibilities. After a couple of years, Menoher gladly took command of the Army's Hawaiian Department.

Pershing brought in Maj. Gen. Mason Patrick, who had successfully managed Mitchell during World War I. Patrick did not put up with Mitchell's theatrics. Once, Billy threatened to resign. Patrick took him to the personnel chief and said, "Go ahead." After that, Mitchell accepted Patrick's authority. Mitchell did well. He was tireless, constantly planning, lobbying, writing and speaking for airpower. He set out to prove that airplanes could destroy warships and take over a critical coastal defense mission, halting an enemy farther from the American coast at a lower cost than a battle fleet, an idea the Navy found hateful.

If people know nothing else about Mitchell, they know that he sank battleships when nobody believed that was possible. The US Navy had quietly tested aerial bombing on an old pre-dreadnought battleship headed for the scrapyard, and it did sink, which they said proved the improbability of a modern ship being sunk from the air. They argued that the old battleship was an outdated, unmanned sitting duck that could not return fire or take damage control measures. Mitchell lobbied Congress for the Navy to provide target ships for a demonstration that would win over the public, perhaps even lead to an air ministry like Britain's.

Mitchell formed the First Provisional Air Brigade at Langley Field in Virginia and had special 2,000-pound bombs made. Sgt. Ulysses Nero, an artilleryman in World War I and a skilled mechanic, invented a bombsight and used it as a bombardier during the demonstration. His success led to a promotion to master sergeant. Nero was later commissioned and retired as a colonel.

The bombers were British Handley-Page O/400s and the newest US bomber, the twin-engine Martin MB-2. The first missions were flown on June 21, 1921, with plenty of brass and press on hand. Captured German ships were the first targets. A U-boat, a destroyer and a cruiser were sent to the bottom of Chesapeake Bay. On July 20, the tests resumed with a battleship, *Ostfriesland*, as the target. One of the best-armored ships in the world, she had fought at Jutland. This was the main event and brought out the Secretaries of War and the Navy, General Pershing, dozens of congressmen and reporters. Mitchell observed from his DH-4B, flying above the scene. The first day's bombing left the battlewagon afloat.

The next day, Mitchell's bombers rained a series of deliberate near misses that crumpled the hull underwater, and soon the Kaiser's dreadnought was gone. The public had seen what bombing could do. The Navy, still decrying results obtained against sitting ducks, built its first carriers.

In 1925, the US Navy dirigible *Shenandoah* crashed in a thunderstorm, killing 14 men, including Mitchell's friend, Lt. Cdr. Zachary Lansdowne. Mitchell told reporters that the accident resulted from the Army and Navy leadership's incompetence and "almost treasonable administration of the national defense." That sort of talk will get you in trouble, even if you are the most famous airman in America. Court-martialed late in 1925, Mitchell was found guilty of conduct prejudicial to good order and discipline. He was sentenced to a five-year suspension. He resigned his commission but spent the rest of his life advocating for airpower. Billy Mitchell died of a heart attack at age 56 in 1936.

Despite this, Mitchell was a prescient observer. In 1924, he predicted that the next war would begin with a Sunday morning air attack on Pearl Harbor by Japan. The Army dismissed that forecast as "exaggerated" and "unsound." On November 29, 1941, the Army-Navy football game program featured a photo of a battleship, captioned, "Despite the claims of air enthusiasts, no battleship has yet been sunk by bombs." The ship pictured was the *USS Arizona*, just eight days away from the Japanese attack at Pearl Harbor.

In 1946, Mitchell was awarded a Congressional Gold Medal, and he is today honored as "the father of the Air Force."

The Air Service in Mitchell's time was obsolescent. Few American-built airplanes got into combat in 1918. The government contracted to build British DeHavilland DH-4 two-seaters, and in the summer of 1918, some arrived on the Western Front, but hundreds were still in the United States. After the war, these leftovers became the Air Service's primary equipment. Jimmy Doolittle made the first transcontinental flight across the US in a DH-4 in September 1922, flying from Jacksonville, Florida to San Diego, California in 21 hours and 19 minutes, with a refueling stop at San Antonio, Texas. Throughout the 1920s, DH-4s were Air Mail carriers for the Post Office. DH-4s were the Air Corps' mainstay till 1932.

Maj. Gen. Mason Patrick was the Chief of the Air Service from 1921 to 1927. A strong advocate for Airmen, he first became involved with the Air Service in World War I. The Air Service organization in France was a mess. Several officers, including Mitchell and Foulois, made a soup sandwich of the Air Service's training, logistics and production. Patrick, a brilliant engineer, was placed in charge. A no-nonsense taskmaster, Patrick put Mitchell in operational leadership, while Foulois was in a headquarters position.

After the war, Patrick returned to the engineers but found himself back in the Air Service after Pershing became chief of staff. Patrick proved to be an excellent airman, adopting many of Mitchell's ideas, but pursuing them "through channels"—that is, in the accepted military manner. Earning his wings at age 59, Patrick became a respected and beloved leader of Airmen. Even today, there is a base named for him.[4] His quiet work to establish technical and engineering facilities left a lasting impact on Airmen as scientific innovators. Mitchell was flamboyant and reveled in publicity. Patrick was quietly effective. Legislation changed the Air Service to an army branch, the Air Corps, in 1926. It remained a small force with outmoded equipment, seeking to stay relevant in a world that had fought "The War to End All Wars."

Record attempts were one way that Airmen stayed before the public. On May 2, 1923, Lts. Oakley Kelly and John Macready took off in a Fokker T-2 from Mitchel Field, New York, and landed in San Diego, California 27 hours later. This was the first nonstop transcontinental flight. A few weeks later, the first aerial refueling occurred on June 23, 1923, when two DH–4s joined up over San Diego, and one aircraft passed gasoline to another through a hose. Extending aircraft range through aerial refueling wouldn't become part of the Air Force mission until the Cold War necessitated intercontinental flights. But in 1923, Lts. Virgil Hine and Frank Seifert refueled Capt. Lowell Smith and 1Lt J.P. Richter. Onloading 75 gallons of fuel, the receiver logged more than six flight hours. A second mission was flown the next day. The receiver remained airborne for almost 24 hours, undergoing several refuelings during this time.

FIRST FLIGHTS AROUND THE WORLD

The most audacious mission yet came next: The first global circumnavigation by airplane occurred when the Army Air Service sent four floatplanes around the world. It took 175 calendar days, totaling 364 flight hours. On April 6, 1924, four **DOUGLAS WORLD CRUISER** seaplanes launched from Lake Washington at Seattle, Washington, flying north to Alaska, along the periphery of Asia—Japan, Shanghai, Hong Kong, French Indochina— then on to Bangkok, Rangoon, Calcutta, Baghdad, Istanbul, Vienna, Paris, London; then Iceland, Greenland, Newfoundland and across the United States. Only two of the original four planes returned to Seattle on September 28, 1924.

Douglas World Cruisers flew around the world

---

[4] Patrick Space Force Base (SFB) in Florida has been an aviation installation since World War II.

The four aircraft were Douglas torpedo bombers, which could be configured with conventional landing gear or as floatplanes. They had extra fuel tanks and were modified for long-range flights.

The planes and crews were: *Seattle* (No. 1): Maj. Frederick L. Martin, pilot and flight commander, and SSgt. Alva L. Harvey, flight mechanic. While Harvey and Martin did not complete the global circumnavigation, their adventure in the wilds of Alaska following a crash into a mountain is a survival story of its own—several days hiking through literally uncharted wilderness, sheltering in a trapper's cabin and then moving downriver to a cannery, where they were rescued. *Chicago* (No. 2) was piloted by Lt. Lowell H. Smith, (of the air refueling experiments in 1923). He became the flight commander when *Seattle* hit that mountain in Alaska. Smith's copilot was Lt. Leslie Arnold. *Chicago* completed the global route and is displayed at the National Air and Space Museum.

The third plane in the flight was *Boston*, flown by First Lt. Leigh P. Wade and SSgt. Henry H. Ogden, flight mechanic. This plane made most of the journey from Seattle across Alaska, the Pacific, Asia and Europe, but had to set down in the Atlantic, north of Scotland. Wade and Ogden were rescued by a US Navy ship, but the *Boston* sank. They got a backup plane in Labrador and finished the mission. Lt. Erik Nelson and Lt. John Harding flew the *New Orleans*, which completed the mission and is now on display at Santa Monica, California.

As aircraft gained capability and range, the search for ways to use air weapons led to the creation of the Air Corps Tactical School (ACTS). It began as a doctrine and tactics course in 1920 and developed into the Air Corps' intellectual center at Maxwell Field, Alabama. This was where high-altitude daylight precision bombing theory was preached as the gospel of the pre-World War II Air Corps. Most air leaders of the era were products of this school. While a few saw fighter, ground attack, close air support and battlefield interdiction missions as vital, most Airmen of that era were enthralled with strategic bombardment as the key to victory.

These evangelists of bombardment were called the "Bomber Mafia." Many of their ideas were implemented during World War II and would shape the Air Force for generations.

A fascinating thought experiment: How might the Second World War and the Cold War have unfolded without this theory? Would aviation have developed in the same way, with intercontinental flight generally replacing intercontinental passenger ships?

Today, Maxwell AFB is the professional education center of the Air Force, from officer training through Air War College, as well as nearby Gunter Air

Force Station[5], the home of enlisted education.

Charles Lindbergh made the first solo nonstop crossing of the Atlantic Ocean in May 1927, sparking unprecedented interest in aviation. Thousands of those who followed his exploits would fly in World War II and be part of the aviation industry that followed. "Lindy" earned his wings as a pilot in the Reserve, graduating at the top of his pilot training class, and was a captain in the Missouri National Guard at the time of his transatlantic flight. He was awarded high military honors for his flight: the Medal of Honor, the Distinguished Flying Cross, and promotion to colonel in the Reserve. The British and French presented him medals. Lindbergh did technical evaluation for the Air Corps in the 1930s and reported on German and Soviet aviation. However, his isolationist views and tolerance of German National Socialism kept him from serving on active military duty in World War II. He was a consultant in aircraft manufacturing and traveled to the South Pacific, teaching fighter pilots how to maximize range through fuel conservation. He flew fifty combat missions and shot down one Japanese airplane. After the war, Lindbergh continued as an adviser to the Air Force and, in 1954, was promoted to brigadier general in the Air Force Reserve.

In June 1927, just a month after Lindbergh's historic flight across the Atlantic, Lts. Lester Maitland and Albert Hegenberger completed the first nonstop flight between California and Hawaii. For the flight to Oahu, Maitland was the pilot, while Hegenberger navigated using dead reckoning, celestial observations and contact with naval vessels. The two were awarded the Mackay Trophy and the Distinguished Flying Cross.

Maj. Carl Spaatz planned an endurance flight that required air refueling to test how long the **Question Mark**, a C-2 Fokker trimotor transport airplane could stay airborne. He rounded up a few top-notch fliers: Lt. Pete Quesada, Maj. Ira Eaker, Lt. Harry Halverson, and a mechanic, Sgt. Roy Hooe. On January 1, 1929, they took off from Van Nuys, California. A

The Question Mark stayed in the air for six days

Douglas C-1 transport would fly over the *Question Mark* every few hours, lowering a 50-foot-long rubber hose to pass gas and send down oil, food and

---

[5] An Air Force Base usually has a runway. An Air Force Station does not.

water via a rope. This continued in the California skies until an engine quit. On January 7, 1929, the *Question Mark* landed after a 150-hour flight with 43 air refuelings that transferred 5,660 gallons of gasoline (or, as tanker crews would say, about 35,000 pounds of fuel offloaded). The crew of the *Question Mark* went on to fine careers in the Air Force. Roy Hooe is the least famous of them. Born in 1892, he was Billy Mitchell's mechanic during the bombing tests and was chief mechanic for the first trans-Pacific flight from California to Hawaii. Hooe retired as a master sergeant in 1950, and died in 1973.

As every "tanker toad" will tell you, this fantastic display of endurance and range would not have been possible without tanker gas. The tankers had the memorable callsigns "Refueling Plane #1" and "Refueling Plane #2," (RP 1 & RP 2). The RP 1 pilots were Capt. Ross G. Hoyt and Lt. Auby C. Strickland, with Lt. Irvin A. Woodring handling the refueling apparatus. RP 2 was flown by Lt. Odas Moon and Lt. Joseph G. Hopkins, and Lt. Andrew F. Solter operated the fuel transfer system. Here's a bit about those tanker crewmen:

Ross Hoyt enlisted in 1914 and was a Balloon Observer in World War I. He became an airplane pilot in 1924 and led a fighter wing in World War II. He retired as a brigadier general in 1945. The Air Force's "Best Air Refueling Crew" award is named for him. Strickland and Hopkins also retired as brigadier generals. Odas Moon died in 1937. Andrew Solter burned to death in a crash in 1936. Irvin Woodring went into flight test at Wright Field and was killed when his aircraft exploded in flight on January 20, 1933, at Wright Field, Ohio.[6] Carl Spaatz once said that his life had been one long series of funerals. Most fliers knew dozens of men who had died in airplanes. Then came combat. Heroes created the Air Force.

The Question Mark crew was awarded the Distinguished Flying Cross (DFC). The tanker crews received a letter in their personnel files. In 1976, the surviving tanker pilots were finally awarded DFCs. Tanker crews, used to toiling in thankless anonymity, smiled wryly at the belated recognition.

INSTRUMENT FLIGHT

Instrument flight was a significant advance. In 1929, Lt. Jimmy Doolittle used a directional gyro, artificial horizon, altimeter and radio navigation to perform an instrument flight, from takeoff to landing, at Mitchel Field, New York. With a safety pilot in the backseat, he flew under a hood. A weapon—or

---

[6] The Davis-Monthan Airfield Register (https://dmairfield.org) presents biographies of aviation personalities who transited that aerodrome from 1925 to 1936. On Lt. Irvin Woodring's page is a copy of his marriage license from 1928. His best man was "Ernest H. Lawson of Selfridge Field, Michigan." Lawson and Woodring were stationed together there. During World War II, Col. Ernest Lawson commanded the 305th Bomb Group and was killed in action, leading the Eighth Air Force on a mission to Hamburg on June 18, 1944.

a mode of transportation—that can only be used in daylight or good weather is not useful. Instrument flight changed that.

William Ocker enlisted in the Army during the Spanish-American War and, by 1909, was a sergeant stationed at Fort Myer, Virginia, where he saw the Wrights' demonstrations and became enamored with flight. He was assigned to aircraft maintenance but earned a pilot's license in 1914. In 1917, he was commissioned and spent much of his career developing flight instrumentation. As a pilot, he understood the problem of flight when vision was obstructed by weather or clouds. As a maintainer, he saw the possibilities of "turn and bank" indicators, to show a plane's orientation relative to the ground. Ocker's work in "blind flight" (instrument flight) laid a foundation for the future Air Force as an all-weather weapon.

The need for this all-weather capability became apparent in the Air Mail debacle of 1934. Federal air mail contract scandals led Secretary of War George Dern to tell President Franklin Roosevelt that the Air Corps could carry mail better than commercial air carriers. When asked about it, Air Corps chief Maj. Gen. Benjamin Foulois didn't give it much thought but said something like, "Sure, the Air Corps could do that." Foulois had committed Airmen to delivering the mail without consulting his boss, Army Chief of Staff Douglas MacArthur. Furthermore, Foulois did not consider that the Air Corps was ill-equipped and poorly trained to fly in instrument conditions and during inclement weather. In February 1934, President Roosevelt ordered the Air Corps to carry the mail immediately.

The Air Corps had plenty of trainers and pursuit planes built to fly in daylight and good weather. Many did not have radios, so radio navigation was impossible. Many pilots assigned to the Air Mail mission had little night or instrument experience. Biplane bombers could carry a hefty payload but in a headwind, they were slower than trains.

Foulois had bitten off more than his Airmen could chew. Less than three weeks into the air mail mission, 10 men had been killed. Eddie Rickenbacker told the press that the deaths were "legalized murder," and Billy Mitchell, never a fan of Foulois, said, "If an Army aviator can't fly a mail route in any sort of weather, what would we do in a war?" The postmaster general reinstated airline air mail contracts. By June 1, 1934, the Air Corps was out of the mail business. Foulois attempted to salvage this situation, stating that it identified Air Corps needs. Nobody bought that, and Foulois retired "for the good of the service" on December 31, 1935. There was none of the usual hoopla—no farewell ceremony, parade, or flyover. But Foulois lived a long life and became a grand old man of the Air Force, who lived to see the advent of space travel. He advocated for airpower and Airmen until he died in 1967.

New aircraft came along in the 1930s. The **MARTIN B-10** entered service in 1933. Its innovations included an internal bomb bay, retractable landing gear, a rotating gun turret, and an enclosed cockpit. It was a giant leap forward from the Keystone and Condor bombers it replaced—giant biplanes that looked like something from the Western Front in 1918. The B-10 resembled early World War II aircraft.

The **BOEING P-26** was introduced in 1932. It was a monoplane, the last pursuit airplane with an open cockpit and fixed landing gear. It could reach speeds of 230 miles per hour and fly 360 miles, but was soon outclassed by designs such as the Messerschmitt Bf-109, Hawker Hurricane and the US-built Republic P-36 and Seversky P-35, all of which had more powerful engines and enclosed cockpits. The P-26 saw combat in the Spanish Civil War, and with the Chinese, US and Philippine forces in 1941.

The **SEVERSKY P-35** came along in 1937. The first all-metal US fighter with an enclosed cockpit and retractable landing gear, its performance would render it obsolete early in the war. A few saw action at Pearl Harbor and in the Philippines. The Swedish Air Force bought the P-35 and flew them until the early 1950s. Strangely enough, the Japanese Navy purchased 20 P-35s and used them in their war with China. There was also a Curtiss P-36 fighter, but it, too, was soon obsolete.

GHQ Air Force (General Headquarters) was initiated under Brig. Gen. Frank M. Andrews in 1935 at Langley Field, Virginia, to control tactical aviation units in the US. This was the first step toward independence, as the operational combat establishment was separated from the training and support forces. General Andrews asserted himself in this role and advocated for long-range bombers and expanding the air mission beyond battlefield support of troops. He was "exiled" to San Antonio in 1939, in the same billet Billy Mitchell had been placed, with a reversion in rank to colonel. Airmen saw it as a punishment, but such a move was not unusual in a small Army where rank was sometimes tied to particular assignments. Fortunately, Gen. George C. Marshall became Army Chief of Staff in September 1939 and brought Andrews back to Washington in a one-star billet as assistant chief of staff for operations. Andrews was named commander of the Caribbean Defense Command in 1941. In January 1943, at the Casablanca Conference, Lieutenant General Andrews was placed in command of US forces in the European Theater of Operations, allowing General Dwight Eisenhower to focus on operations in North Africa and Italy. On May 3, 1943, Andrews was killed in a crash in Iceland. Andrews AFB is named for him.

THE BOEING B-17 FLYING FORTRESS was the embodiment of the High Altitude Precision Daylight Bombardment theory. A four-engine heavy bomber aircraft developed in the 1930s, Boeing's Model 299 prototype was what the Air Corps wanted, but after it crashed, the War Department chose cheaper two-engine Douglas B-18 Bolo bombers. The Air Corps ordered 13 B-17s for evaluation, which came into service in 1937. The opportunity to showcase the big bomber came with the *Rex* Mission in May 1938. This publicity venture involved sending

B-17s to intercept the Italian ocean liner *Rex* as it approached New York. The B-17 flight's leader was Maj. C.V. Haynes, and the lead navigator was Lt. Curtis LeMay. They took off from historic Mitchel Field, near Queens, New York, where Kelly and Macready departed on the first transcontinental flight across the US, and where Doolittle did the first "blind" flight. It would also become the Army Air Forces (AAF) Antisubmarine Command headquarters in World War II. It was named for former New York Mayor John P. Mitchel, who was killed in a flight training accident during World War I.

The three B-17s found the ship more than 700 miles from New York, despite driving rain and an overcast that made sighting the ship difficult. Air Corps Publicist Lt. Col. Ira Eaker had invited reporters and arranged a live nationwide radio broadcast from the lead bomber as they made the intercept, on time. Photographers and movie cameras were also aboard to capture this Air Corps triumph. The Navy was unhappy with the Airmen extending their reach on the high seas. In a few years, Airpower's reach was clear to all.

CHAPTER 4
WORLD WAR II

"You've got to kill people and when you kill enough of them, they stop fighting." - Curtis LeMay

It seems strange to say, but World War II was part of my childhood, though it ended more than a decade before my birth. Before Vietnam, the military was held in the highest esteem, perhaps because its most significant achievement was fresh in memory and so many people had participated in that victory. There were lots of military-themed TV dramas like *Combat!* and *Twelve O'Clock High*. There was a documentary series called *Battle Line* in which a veteran from each side would describe what he saw—Americans, British, German and Japanese veterans, each telling his side of the story. We watched *Victory At Sea* on Sundays, the US Navy steaming across the Pacific and sweeping the U-boats from the Atlantic, all set to Richard Rodgers's Broadway show music. The Army had a documentary series called *The Big Picture*. There was a series called *The Twentieth Century* that featured many war stories. Even in 1974, the TV series *The World at War* featured interviews with many senior leaders of World War II. All that surely influenced my interest in all things military.

Now I realize that in my childhood, the World War II vets were mostly in their forties and fifties, in the prime of their lives. Many of my friends' dads, our coaches, teachers and neighbors were veterans. Men whom I saw every day had been in "the war." Listening to them talk about their adventures in the Philippines or Europe, it all sounded like great fun, when maybe after a couple of beers they told and retold funny stories about swiping food and booze from the officers' mess or pranks played in camp. I listened avidly to their stories. They had seen places that I couldn't imagine visiting—the South Pacific, the Philippines, Australia, North Africa, Europe... it all seemed grand and glorious to serve in such a noble cause.

But war meant other things, horrible things. There were a couple of uncles killed in the war and Uncle Bill never talked about what nightmares he had seen on Saipan and Iwo Jima. Rarely, those dads and uncles and coaches might talk among themselves about darker moments: "This kid put that flame thrower right on those Japs..." "Our platoon sergeant shot that German lieutenant and pulled a map out of that Kraut's pocket..."

I was about eight years old the first time I saw photos of Holocaust victims, in my grandpa's copy of the *American Heritage Picture History of World War II*. Emaciated corpses with dead, staring eyes scared me. Still, as I learned of brave men who defeated the Nazis, I came to realize that those who stopped this great evil had done something honorable and righteous.

Some of those heroes were American Airmen.

The United States Air Force has had many names. From 1907 to 1914, it was the Aeronautical Division, Signal Corps. Until May 1918, it was the Aviation Section of the Signal Corps. It was the Division of Military Aeronautics for a few days in May 1918, before it became the Army Air Service. Legislation then created the Army Air Corps, as it was known from July 2, 1926, to June 20, 1941. From 1941 to September 18, 1947, the organization was known as the United States Army Air Forces. This made the AAF equal with the Army Ground Forces (AGF—the combat arms) and the Army Service Forces (ASF—supporting elements). Its first commander was Hap Arnold.

Henry H. "Hap" Arnold is revered by Airmen but little known outside the Air Force. He was not a combat leader nor a tactical genius. Leading and building the most powerful air force in history, he directed global operations, research and development, procurement and everything in between. He worked to gain Air Force independence, and in this he won the support of General George Marshall. His nickname came from youthful pranks and a ready smile, but he could be a ruthless taskmaster. He had four heart attacks during the war and probably should have been medically retired, but he continued driving himself and his staff at an inhuman pace. One officer actually dropped dead in front of Arnold's desk.

Henry H. "Hap" Arnold, 1911

Arnold was a West Point graduate and one of the first pilots. During World War I, he was primarily assigned to Washington, the youngest colonel in the Army at 31. His experience in procurement, building training facilities, operating airfields and working in the capital served him well in World War II. Arnold was an acolyte of Billy Mitchell and, like Mitchell, a talented writer. Arnold demonstrated how airpower fit into a combined-arms team, and in 1934 he led a highly publicized flight of B-10s to Alaska. After the death of Maj. Gen. Oscar Westover in 1938, Arnold became Chief of the Air Corps. Arnold's World War II AAF was the largest air force in history. He received his fifth star in December 1944, the only Airman to wear that highest rank. He left active duty in 1946 and died following a fifth heart attack in 1950. Hap Arnold's legacies are many, including the officers who caught his eye—Carl Spaatz, Hoyt Vandenberg, Nathan Twining, Lauris Norstad and Curtis LeMay—who led the Air Force through the 1950s.

The AAF's plan to defeat Germany was AWPD-1, developed by the Air War Plans Division in 1941. Members of the "Bomber Mafia"—Lt. Cols. George and Walker, and Majs. Hansell and Kuter—planned a strategic bombing campaign. They believed that wrecking electric power, transportation, petroleum and the German Luftwaffe (air force) would win the

General Henry H. "Hap" Arnold, 1945

war. This was a dedicated effort to determine what industrial, economic and military systems would have to be destroyed to win the war. While imperfect, its estimates provided a blueprint for war planning. It was modified as political and tactical considerations emerged. For example, submarine pens rose in priority as the Battle of the Atlantic threatened Britain. The planners came close to estimating the forces required for victory. AWPD-1 foresaw 2.1 million Airmen and 63,000 aircraft, including B-29 and B-36 bombers, to win the war. The actual numbers were 2.4 million AAF men and nearly 80,000 aircraft. The B-29 was not used in Europe, and the B-36 did not appear during World War II.

Bombers were the backbone of the plans. From its birth in 1935, the B-17 Flying Fortress promised to marry bombing theory—high altitude daylight precision bombardment—with the range and self-protection needed to make the theory a reality. In 1938, Lt. Col. Robert Olds led a flight of six B-17s on a goodwill mission to Argentina, as the Air Corps showcased the B-17 in highly visible missions. The Flying Fortress underwent many modifications during the war, but more than 12,000 were built.

The second bomber developed was the **CONSOLIDATED B-24 LIBERATOR**, which entered the AAF inventory in June 1941. More than 18,000 B-24s were produced during World War II, more than any other US aircraft. The beautiful B-17 has always gotten more publicity, but the B-24 carried a heavier bomb load, was faster and had more range. But it was not as sturdy as the aptly named Flying Fortress. The B-17 was more heavily armed, could take more battle damage, and could fly higher. The B-24 served well as a bomber, notably in the Mediterranean and Italy, and its long range made it well-suited to the Pacific and to antisubmarine patrols. Some were even adapted to fly cargo missions. The Liberator's drawbacks? It tended to catch fire when hit, was difficult to bail out of, and could easily lose its wings if hit in the right spots.

The two bombers' crews had a rivalry that a small item in the flight manuals illustrates: The B-17 manual's go-around procedure—an aborted landing—had a sketch of a wrecked Liberator on the runway. The B-24 manual depicted a busted-up Fort. Flight crews are like families—they can make fun of each other, but outsiders are best advised to stay out of it!

On Sunday, December 7, 1941, about 400 AAF aircraft were stationed in Hawaii. In the attack that day, 188 were destroyed and 150 were damaged. AAF pilots shot down 10 Japanese planes. Lts. George Welch and Ken Taylor shot down four apiece and received the Distinguished Service Cross. Welch ended the war as an ace with 16 kills in the Pacific but died in 1954 while flight testing the F-100 Super Sabre. Taylor served on Guadalcanal and was severely wounded in a Japanese air raid in 1943. He retired as a brigadier general and died in 2006.

The first Americans to participate in the European air war were volunteers with the Eagle Squadrons, fighter units comprised of American volunteers with the British Royal Air Force, similar to the Lafayette Escadrille of the First World War. Formed in 1940, a few hundred Americans served in the Eagle Squadrons. About 100 were killed or captured. Most of the Eagles transferred to the USAAF in 1942, as American units entered the war. Among them were aces Chesley Peterson, Don Blakeslee and Don Gentile. Peterson had been eliminated from Air Corps pilot training in 1939 but won his wings with the RAF. He flew 42 missions and won several British decorations before transferring to US service and becoming the youngest American colonel at 23. Blakeslee was an ace with the RAF, then joined the US forces in England. He flew at least 500 sorties, with more than 1,000 hours of combat time; often, he would not log his own time and would instead credit other pilots with his aerial victories. He led the first American fighters over Berlin and flew the first shuttle mission—from Britain over Germany, landing in the USSR. His 4th Fighter Group scored the most victories (over 1,000) in air-to-air combat. Gentile was a household name during the war as one of the top aces. He died in a flight accident in 1951.

Ira Eaker was the commander of the AAF bombers based in England. Born in 1896, he entered the Army in 1917 and earned his wings in late 1918. He accomplished the first transcontinental flight entirely on instruments in 1930. He was the publicist for the *Rex* mission in 1938.

In December 1942, Eaker took command of the Eighth Air Force. He worked tirelessly for daylight "precision" bombing of enemy military and industrial targets at a time when the RAF said this was impossible, so they were bombing at night. When Winston Churchill read Eaker's words, "round the clock bombing," he savored them, saying the "devils shall get no rest." Eaker flew the first US bomber mission against German targets in France on August 17, 1942. Bombing Germany was hindered by heavy losses to a still-building bomber force, primarily due to a lack of fighter escort. As the war in Europe ended, he returned to the United States as deputy commander of the Army Air Forces and Chief of the Air Staff. He retired in 1947. He became an executive in the aerospace industry and a newspaper columnist. Ira Eaker died in 1987.

Glenn Miller was one of the top bandleaders at the outbreak of World War II, with more than 60 Top 10 hits. At the time of Pearl Harbor, he was 37 years old, a married man with two kids and bad eyesight. He was earning $20,000 a week, astronomical money in 1941. Still, he sought a commission and entered the Army as a captain in October 1942. He was assigned to lead the AAF Training Command Band. The motto of the AAF Technical Training Command—"I Sustain the Wings"—became a hit and the theme song of a radio program that aired from June 1943 to November 1945. Miller assembled a superb collection of big band, jazz and symphony musicians, and in the summer of 1944, his band was assigned to the European Theater of Operations. They performed for troops and the British people, and their music was broadcast on the radio in the United States, Britain and, as a propaganda move, Germany. Miller's music drew German listeners, and the program could tell Germans that the war was a lost cause and that the world would be a better place after Hitler. With a German-speaking hostess and Miller speaking from a phonetic German script, it was an underground hit in Germany.

While preparing to move the band from England to France, Miller was lost when the aircraft in which he was traveling went missing in the English Channel in December 1944. No trace of the plane or its occupants has ever been found.

TOP ACES

Military aviation enthusiasts can name the top aces of World War II. Names like Dick Bong, Gabby Gabreski and Tommy McGuire are well-known. But what about the other guys shooting at German fighters as if (and because) their lives depended on it? The gunners in bombers shot down thousands of enemy planes, but they are not so widely remembered. Part of that is due to the lone warrior mystique surrounding fighter pilots, the single combatant engaging in knightly combat against a worthy foe. On the other hand, bomber gunners were part of a team, hauling a bombload to the enemy. The gunners were typically enlisted, lower-ranking crew members, toiling anonymously amid hundreds of other bombers. There is also the problem of determining who shot whom: Several gunners might fire on a given fighter, making it difficult to say who got the kill shot. Perhaps 300 aerial gunners were credited with shooting down five enemy airplanes each.

A few gunners received special recognition during the war. Master Sergeant Michael Arooth was officially credited with nine victories but may have downed 17 German fighters in just 14 combat missions during the summer

Maynard "Snuffy" Smith, bomber gunner, 1943

of 1943. A tail gunner in the 379th Bomb Group, Arooth was awarded a Distinguished Service Cross for a mission against Kassel, in which he shot down four Germans despite severe wounds. He also received two DFCs, four Air Medals, and two Purple Hearts. He later suffered a head injury during a ditching in the English Channel, which ended his flying career.

Staff Sergeant Donald Crossley, a B-17 tail gunner, shot down 12 Germans. Crossley flew in the 95th Bomb Group and appeared on the cover of the *New York Times Magazine* in October 1943.

Technical Sergeant Arthur Benko, a B-24 top turret gunner of the 308th Bomb Group of the 14th Air Force in the China-Burma-India (CBI) theater, was credited with 16 kills. A noted marksman in Arizona, he fought a tremendous battle against Japanese fighters in a raid at Haiphong in Vietnam. On October 1, 1943, his squadron was the last over the target and was jumped by waves of enemy fighters. The running battle lasted nearly an hour, with Benko downing seven Japanese Zeros despite wounds to his head and wrist. A few weeks later, returning from a mission to Hong Kong, Benko's plane developed engine trouble and the crew had to bail out over China. Many landed in friendly territory, but Benko remains listed as Missing in Action (MIA).

Two famous gunners were TSgt. John Quinlan, who shot down five enemy fighters—two over Europe as a tail gunner on the celebrated *Memphis Belle*, and three more as a B-29 tail gunner; and SSgt. Benjamin Warmer, a B-17 waist gunner in the 99th Bomb Group, 12th Air Force. During a bombing mission to a target in Sicily on July 5, 1943, Warmer shot down seven German fighters, for which he received the Distinguished Service Cross. He finished the war with nine enemy planes destroyed.

Another famous gunner was Lee "Shorty" Gordon, the first American to escape from a German prisoner of war (POW) camp and make it back to friendly territory. A ball turret gunner in the 305th Bomb Group, he was shot down on February 26, 1943 near Wilhelmshaven. Taken prisoner, he immediately began plotting an escape. He bribed guards with American cigarettes and coffee obtained from Red Cross packages. Then, the five-foot-two sergeant made his getaway. After a four-month trek across France, with the help of the French Underground he made it back to England a year and a day after he left.

There were bomber gunners in Korea and Vietnam, and the gunners of the B-29 *Command Decision* were credited with destroying five enemy aircraft as a crew. The last two bomber gunners to shoot down enemy fighter planes were SSgt. Samuel O. Turner and A1C Albert Moore, B-52 tail gunners who downed MiGs near Hanoi during Linebacker II in December 1972.

NAVIGATORS

Before 1940, a handful of aerial navigators were employed by Pan American Airlines for their Clipper routes across the Atlantic and Pacific. A few Air Corps pilots had learned the arcane skills of aerial navigation. So, when the war came and large aircraft—bombers and transports—began crossing the oceans and uncharted land, there was a need for men who could find their way over the earth. Using techniques from thousands of years of sea navigation—dead reckoning (I'm on 090 heading at 200 knots and have been doing this for two hours, therefore I must be at a point 400 miles east of my last known position, plus the effect of wind); pilotage (observing landmarks); and celestial sightings (sun, moon, Venus, Mars, Jupiter, Saturn, and stars), the AAF produced 50,000 navigators during the war. Many of these young men had never flown before entering training or been out of sight of land before they moved out for the war zones. Most of them made it across oceans and distant continents.

Many student navigators had not flown before the war

Still, throughout World War II, approximately 1,000 AAF planes vanished—just disappeared. That is about five per week throughout the war. While the enemy may have downed some, many just got lost or suffered mechanical malfunctions. In many cases, we don't know what happened.

Eddie Rickenbacker and Maj. Gen. Nathan F. Twining were two of the most famous aviators to survive ditching in the Pacific. In October 1942, Rickenbacker was traveling to the South Pacific for the War Department when his B-17 ran out of fuel over the ocean. He and six other men spent more than three weeks in a life raft before being rescued. It is said that after that, he kept a pitcher of water beside his bed every night, and by morning, he had drunk it all. Twining spent six days in a raft with 14 others after they went down in the Pacific in February 1943. Medal of Honor recipient Leon Vance, band leader Glenn Miller and Generals Millard Harmon and James Andersen were aboard airplanes that disappeared without a trace at sea. Fortunately, better navigation systems evolved. Long Range Navigation (LORAN) uses radio waves to measure distances to a station. It came into use during the war. Inertial Navigation Systems (INS) were in use by 1970. In the 1990s, GPS ended the need for a navigator with a table full of maps, navigational charts, almanacs, star diagrams and a sextant.

UNOFFICIAL SURVIVOR CLUBS

Those who survived bailouts, ditching, or escaped captivity often gained some unofficial recognition. Here are three of the most famous:

Caterpillar Club—Sponsored by companies that made parachutes. To join, one must have survived a disabled airplane by using a parachute (which was made of silk). Occasionally, the caterpillar pins are seen on old uniforms.

Goldfish Club—Those who survived a wartime ditching or bailout. Their embroidered winged goldfish on the waves patch was often sewn under a dress uniform pocket flap.

Winged Boot Club—This insignia is worn by those who escaped and evaded through enemy territory to friendly lines. This tradition began with British Airmen in the North African desert, then was picked up by American Airmen who successfully escaped Nazi Europe. The winged boot symbolizes flying into the hostile zone and walking home from it. Like other unofficial insignia, it was usually hidden under a pocket flap or a lapel but occasionally could be spotted out in the open.

Even today, those who eject from aircraft using Martin-Baker ejection seats receive small gifts from the company.

NOTEWORTHY

FOURTH FIGHTER GROUP: The highest-scoring fighter group of the war was the 4th Fighter Group. Led throughout much of the war by brilliant fighter tactician Don Blakeslee, the 4th began primarily with Americans from the Eagle Squadrons. They initially flew Spitfires, then transitioned to P-47s, and later to P-51s. The 4th was the first American fighter unit over Paris, Germany and Berlin. They destroyed 1,016 enemy aircraft, and 81 pilots became aces, including such heroes as Don Gentile, Duane Beeson, John Godfrey, James Goodson, Ralph Hofer and Vermont Garrison.

They lost 241 aircraft; of the 555 pilots who flew with the unit throughout the war, 128 were killed and 105 captured. No matter how good fighter pilots might be, a little bad luck or a momentary deviation could be deadly.

TUSKEGEE AIRMEN: The 99th Pursuit Squadron was activated on March 22, 1941, at Chanute Field, Illinois, to train black pilots. In July 1941, the first class of black pilot candidates in the Army Air Forces reported to Tuskegee, Alabama. The first Tuskegee Airmen received their wings on March 6, 1942. Capt. Benjamin O. Davis Jr. was one of them. He would command the Tuskegee Airmen in combat. In World War II, 992 pilots were Tuskegee Airmen. Of these, 355 went overseas, 84 died, 68 were killed in action, 16 died in mishaps and 32 were taken prisoner. They destroyed 112 enemy aircraft in the air, 150 on the ground and 1,000 railroad engines and cars.

Lt. Charles Hall became the first black AAF pilot to shoot down an enemy airplane, a FW-190, over Sicily, on October 2, 1943. Hall also became the first African American fighter pilot to earn the Distinguished Flying Cross. During his 198 combat missions, he downed three German FW-190s. Late in 1944, Hall returned to the US for a war bond tour and served as a flight instructor at Tuskegee.

Benjamin O. Davis Jr. earned a Silver Star and a DFC, and became the first black general in the Air Force. His father was the first black general in the Army and was the only black officer in the Army until his son was commissioned at West Point in 1936. The younger Davis led the Tuskegee Airmen throughout the war and became the group commander. After the war, he continued to rise, pinning on his first star in 1954. He retired in 1970 as a lieutenant general. He was awarded a fourth star in retirement and passed away in 2002.

WOMEN FLIERS IN WORLD WAR II: Nancy Harkness Love was a member of a wealthy Philadelphia family and learned to fly in the 1930s while attending Vassar. She married Robert Love in 1936; he was an Air Corps Reserve officer who was a leading figure in the Air Transport Command in World War II and ran Allegheny Airlines after the war.

In 1942, Robert mentioned his wife's aviation skills to his boss, Col. William Tunner, who realized female pilots were an untapped resource. Nancy Love was named head of the Women's Auxiliary Ferrying Squadron (WAFS), which began delivering aircraft from factories to bases and to overseas departure points in September 1942. Jacqueline Cochran, a famous aviatrix, ran a female flying unit, the Women's Flying Training Detachment (WFTD), doing similar work. With overlapping missions, the two groups merged as the Women Airforce Service Pilots (WASP) in August 1943. More than 1,000 women served, ferrying aircraft, flying cargo and passengers, towing targets and testing aircraft. The WASPs disbanded in December 1944. In 1977, the WASPs were granted military status.

Carl A. Spaatz was the last Commanding General of the Army Air Forces and the first Chief of Staff of the Air Force (CSAF). Few men were as impactful in so many ways: combat hero, airpower advocate, bomber leader, then a founder of the Air Force. Nicknamed "Tooey," he was a fighter pilot in World War I. Spaatz shot down three Germans and earned a Distinguished Service Cross.

A friendly man, quick to laugh, strum a guitar, deal a hand of cards or share a drink, Tooey didn't excel academically but was an operational wizard, a staunch believer in offensive airpower and bombardment.

In 1929, he led the *Question Mark* flight. He was an observer during the Battle of Britain, then commanded bomber operations. In tactical roles in North Africa and as commander of Strategic Air Forces in Europe in 1944-45, his bombers helped roll back the Afrika Korps, paralyzed the German Army in Western Europe and wrecked the German war effort. Biographer David Mets wrote that it was not easy "to think of any other commander who had both the perception to identify oil targets as decisive and the strength to conserve a part of the US strategic air striking power for them."

General Carl Spaatz

General Eisenhower said Spaatz was one of the two American general officers who contributed the most to the victory in Europe. The other was Omar Bradley. Spaatz was the only American present for all three surrenders: the German surrender at Reims on May 7, 1945; at Berlin the next day, when Germany surrendered to the Red Army; and on the battleship Missouri in Tokyo Bay when the Japanese surrendered on September 2, 1945. At the war's end, he replaced Hap Arnold and became the first CSAF in 1947. He retired in 1948 but remained an advocate for airpower. He died in 1974.

WORLD WAR II: EUROPEAN CAMPAIGNS

"I'll never know how many people I helped to kill."
- Bert Stiles, B-17 pilot. Killed In Action, November 26, 1944

The TV series *Masters of the Air*, whose title was taken from Donald L. Miller's book on the Eighth Air Force in World War II, dramatically brought this story to the public. Their heroism in the face of terrifying missions and scant odds of survival remains awe-inspiring. Some 350,000 Airmen served with the Eighth Air Force in England. About 210,000 of them were aircrew. The Eighth's casualties included 26,000 killed and 21,000 captured—nearly one in four crewmen were lost in battle. The Eighth Air Force lost more men than did the entire Marine Corps. On the Schweinfurt-Regensburg raid in August 1943, 60 B-17s were shot down. That was one-sixth of the 376 bombers dispatched, and meant 600 fresh crewmen were needed. In 1942-43, the odds were against bomber crews completing a 25-mission tour in the Eighth Air Force. Only 35 percent survived their tours. As the odds improved and the campaign intensified in 1944, the required missions increased to 30 and then 35, and about 66 percent of the crewmen completed their tours. By 1945, 81 percent of the combatants completed thirty-five missions.

The AAF began its air campaign in Europe on July 4, 1942, as six Boston light bombers, borrowed from the RAF, hit targets in the Netherlands. Those planes were called A-20 Havocs in American service. That day, they were flown by Yanks from the 15th Bombardment Squadron, newly arrived in England. The RAF planes were quickly given US insignia for the sortie. Two of the six were shot down, and Capt. Charles Kegelman, the formation leader, barely got back to England. Hit by enemy fire, he flew so low over Holland that tree branches scratched up the plane's belly. Reading the after-action report, General Eisenhower asked, "Are all the reports going to be like this?"

Awarded a Distinguished Service Cross, Kegelman completed his first combat tour and made colonel at age 29. He died in a B-25 in the Philippines in 1945.

The "Mighty Eighth's" heavy bombers entered combat with a raid on rail yards at Rouen, France, on August 17, 1942. Twelve B-17s under Col. Frank A. Armstrong Jr., along with Maj. Paul Tibbets and Maj. Gen. Ira Eaker, hit their targets and returned safely. For the next several months, the Eighth bombed targets in France and the Low Countries. The naval base at Wilhelmshaven was the first US target in Germany. Only 91 bombers launched on the mission on January 27, 1943, and just 58 reached their targets. They dropped about 137 tons of munitions to destroy U-boat facilities. Three bombers were lost. The raids—and the losses—soon got bigger.

The Fifteenth Air Force flew its first mission on November 2, 1943, sending about 100 bombers against aircraft factories in Wiener Neustadt, Austria. Maj. Gen. Jimmy Doolittle was its first commander when the Fifteenth flew from North Africa and advanced to southern Italy, near Foggia.

Even those short missions were deadly. The Luftwaffe had a strong presence in France, and Fortress Europe was well-defended. Sometimes, survival was just a stroke of fate. Staff Sergeant Alan Magee was a ball turret gunner in the 303rd Bomb Group on a mission to St. Nazaire, France, on January 3, 1943. The target was torpedo storage for U-boats. Over the target, Magee's plane was hit by several flak bursts, went into a spin, and he was thrown from the B-17. Wounded and without a parachute, Magee fell 20,000 feet, hitting the glass roof of the train station, which apparently had enough give in it to break his fall before the glass shattered. He ended up in the structure itself, recovered by French workers, who were shocked to find him alive. Badly cut up, with two dozen puncture wounds and numerous broken bones, Magee was cared for in a German hospital and recovered sufficiently to be sent to a POW camp, where he spent the rest of the war.

The *Memphis Belle* was a B-17F and the subject of a 1944 propaganda film. The Eighth Air Force was battling to carry out the bombing offensive in Europe in early 1943. Despite glamorous images and high expectations, losses were terrible, and results were poor. A film promoting the completion of a crew's combat tour promised to be popular and to reflect well on the AAF. Hollywood film crews and Academy Award-winning director William Wyler flew combat missions to make the film, and cameraman Harold Tennenbaum was killed in action. Wyler's combat footage depicted the air war more realistically than people back home had ever seen it. Capt. Bob Morgan and his crew went on a war bond tour across the United States in 1943. The *Memphis Belle* was kept for many years in Memphis, and is today displayed at the National Museum of the Air Force.

Why the *Memphis Belle*? It was credited as the first US bomber to complete a 25-mission combat tour in the Eighth Air Force. Well, it was one of the first. A few bombers had completed combat tours in early 1943 in the Pacific and North Africa. Still, at this point, it was rare for a heavy bomber to survive 25 missions over Nazi-occupied Europe. Few crewmen were surviving their combat tours. Other aircraft were considered. A B-24 completed its missions, but the Liberator wasn't as beautiful as the Flying Fortress. A B-17 bomber nicknamed *Hell's Angels* finished its tour a few days before the *Belle*—but that

B-17 crew, 305th Bomb Group

381st Bomb Group B-17s in formation

name didn't quite fit the wholesome image desired in 1940s America. William Wyler's crews flew on B-17s, thus their focus fell on a few men, notably the crew of the *Memphis Belle*.

A significant task the AAF faced was putting enough aircraft and crewmen into the fight. With war all over the world and long training pipelines, the AAF had a job just getting itself geared up. By July 1943, the Eighth Air Force could finally send about 300 bombers out, day after day, against Germany. During Blitz Week, July 24-30, 1943, the Eighth struck on six of seven days. Two attacks on Hamburg hit shipyards and an aircraft engine plant during the RAF's Operation Gomorrah, which devastated that port city, killing 40,000 people. While precision bombing was not as accurate as the AAF hoped, it should be noted that the two American raids killed about 170 civilians, far less than died under the RAF's strategy of "area bombing." About 1,000 American Airmen were killed, wounded or captured in that week of attacks.

A key target was the oil refineries at Ploesti, Romania. They produced 30 percent of Axis oil, including most of the Luftwaffe's aviation fuel. Ploesti's importance was apparent. Col. Harry Halverson led 13 B-24s (called HALPRO) to bomb the refineries. HALPRO launched from Egypt and attacked Ploesti on June 12, 1942. The raid caused minor damage but alerted the Germans to the need for upgrades to Ploesti's defenses.

The most famous Ploesti raid occurred on August 1, 1943, during Operation Tidal Wave. Five B-24 Groups flew from Libya to bomb Ploesti. The bombers flew at low level, hoping to avoid detection, just a couple hundred feet above the ground; so close, gunners in the bombers could see Germans manning flak cannons. There were 178 B-24s assigned. One plane crashed on takeoff, and then, approaching the target, a navigational error alerted the defenders to the attacking force. The Liberators roared toward an enemy who was ready for them. Why did this go so badly? Some say the plane that crashed on takeoff carried the lead navigator. Some say the lead planner, Col. Jacob Smart, didn't know enough about the B-24 and low-level flight. Others blame Col. Keith Compton, commander of the 376th Bomb Group, and say he didn't coordinate his flight profile with the other groups (they didn't all launch from the same airfield), and the attack's timing was badly affected. More than 600 Americans were casualties as 54 of the B-24s were lost. Although the refinery was severely damaged, it was repairable. Twenty more missions to Ploesti were flown in 1944. More than 5,600 sorties dropped 13,000 tons of bombs at a cost of more than 220 heavy bombers and 52 medium bombers and fighters. Nearly 1,700 American Airmen were killed, and 1,000 captured. But Ploesti ceased to be a productive oil source for the Germans only when Russian troops overran the area late in 1944.

Five fliers on the August 1, 1943 raid earned the Medal of Honor. The B-24 piloted by Lt. Col. Addison Baker and Maj. John Jerstad was hit by flak three miles from the target. As the 93rd Bomb Group leader, aborting would have possibly led the group to follow them away from the hellscape that was the target. Baker and Jerstad pressed the attack, dropped their bombs, then their plane slammed into the ground, killing all aboard. Baker and Jerstad received the Medal of Honor posthumously. Col. John Kane of the 98th Bomb Group and Col. Leon Johnson, commander of the 44th Bomb Group, led their bombers toward the refineries. Through a storm of flak, the Liberators drove on to their objectives. Kane and Johnson were awarded the Medal of Honor for heroic leadership. The fifth Medal of Honor went to 2nd Lt. Lloyd Hughes. His plane was hit several times, fighting toward the refineries. He continued to the target, one wing completely afire. He tried to set the plane down, but it cartwheeled. He died, but two men survived due to his heroism.

The Schweinfurt-Regensburg raid took place on August 17, 1943. Planners seeking a way to break the back of the Nazi war machine thought destruction of the German ball-bearing industry was promising: Most engines and machines required some bearings and, theoretically, this would damage every aspect of the German war industry. A dual-purpose mission from England looked promising, 230 bombers going to Schweinfurt to hit a ball-bearing factory, while 146 struck an aircraft factory at Regensburg. The expectation was that the Germans would have to divide their fighter force, giving both bomber fleets a better chance to survive. The Regensburg force got airborne on time, with Col. Curtis LeMay in the lead ship. The Schweinfurt package was three hours late due to weather in England. This timing gave the Luftwaffe time to attack the Regensburg force, then refuel, rearm, and hit the Schweinfurt force on its way in—and again on their way home. The Regensburg force was headed for North Africa in another piece of clever planning that did not work out. They drew little interest from the Germans and faced scant opposition on their way south while the defenders focused on battering the Schweinfurt bombers. The two groups totaled 376 bombers, and 60 were shot down that day. The Messerschmitt factory in Regensburg was severely damaged, but Schweinfurt kept turning out ball bearings.

In the fall of 1943, the doctrine of strategic bombing was failing. The Eighth Air Force was taking heavy losses, and targets were not being destroyed. Missions to Stuttgart, Bremen and Munster all suffered losses of 10 percent or higher in September and October. The Eighth returned to Schweinfurt on October 14. That mission saw 60 bombers downed of the 291 planes sent out, making a total of 150 bombers and 1,500 men lost in six days. They did some damage to the factories, but nothing that the Germans couldn't fix. In the fall of 1943, the Eighth had to back off from deep raids into Germany. It was just too costly without fighter escort.

The B-17 and B-24 were capable bombers. But to be effective, fighter escorts had to be able to fly deep into Germany. The P–47 Thunderbolt was a high-altitude escort fighter and a low-level ground attack aircraft. In April 1943, the P-47 went into combat in Western Europe. Rugged, able to take severe battle damage, and to carry serious firepower—six .50 caliber machine guns, an array of rockets, and more than 2,000 pounds of bombs—it was a good airplane. More than 15,000 Thunderbolts were built. But the P-47 could barely get to Germany. Enemy fighters lurked beyond, waiting to pounce on the bombers when their "Little Friends" in the escort had to go home. The P-51 Mustang entered combat on December 5, 1943. By V-E Day, 14 of 15 fighter groups in the ETO were flying Mustangs. Equipped with drop tanks—fuel tanks that could be jettisoned when empty—the Mustang could reach almost all of the Third Reich. P-51s destroyed nearly 5,000 German aircraft. It was the best propeller-driven fighter of World War II. Without it, getting bombers deep into Germany would have remained prohibitively deadly.

HEROIC AIRMEN

The valor of the bomber crewmen is unforgettable. Lt. Col. Manny Klette flew 91 missions against Nazi Germany, in combat from March 1943 to the end of the war. Thirty of these missions were as leader of his group, wing or the entire Eighth. He received the DSC and four DFCs. Lt. Col. Robert Rosenthal flew 52 missions in the 100th Bomb Group. Shot down twice, wounded twice, "Rosie" earned a DSC and two Silver Stars. TSgt. Sator "Sandy" Sanchez was the only enlisted man to have a B-17 named for him, *Smilin' Sandy Sanchez*. He was shot down on his 66th mission over Europe. He flew 44 missions with the Eighth Air Force, returned home, then went to fight in the Fifteenth Air

Force from Italy. He was awarded a Silver Star, DFC and the Soldier's Medal. He was killed in action on March 15, 1945.

On November 29, 1943, a B-17 was hit by flak over Bremen, Germany. Ten of eleven crewmen were hit. Tail gunner SSgt. Joseph Sawicki suffered severe abdominal wounds and his left arm was gone below the elbow. He crawled forward to the waist section where both gunners lay bleeding and dazed. Sawicki buckled their parachutes and pushed them out of the hatch. They survived as POWs. SSgt. Sawicki collapsed and died, his body remained in the plane. Only four of the crew survived the war. After their liberation, the survivors shared stories of that day, and the aircraft commander, Capt. Carl Fyler, wrote Sawicki up for the Medal of Honor. Other crewmen followed up on this, but both recommendations were lost. In 1995, the Army rebuffed further efforts because, by then, there were no living witnesses.[7]

TSgt. Forrest Vosler was a radio operator on a B-17. Over Bremen, the plane was hit by enemy fighters. Vosler was wounded in the legs and face—he was virtually blinded by fragments in his eyes. Despite this, he kept firing at enemy planes, repaired the damaged radio, and sent out distress signals as ditching in the North Sea became imminent. Hitting the water, Vosler held onto the wounded tail gunner until other crewmen could assist. Vosler lost sight in one eye and the other never fully recovered. He received the Medal of Honor.

It is often noted that German war production stayed strong until the last months of the war. About one million men fought in the defense of the Reich, using thousands of artillery pieces. Without the bomber offensive, these could have been used at the fighting fronts. Could this have changed the outcome? Doubtful. Could it have made victory far more costly? Certainly, production would have soared without the constant destruction of aerial bombardment.

Another factor in the Mighty Eighth's success was that Jimmy Doolittle took command of Eighth Air Force in January 1944. Doolittle had worked with Dwight Eisenhower in North Africa. Ike's first impression of the famed racing pilot and Tokyo Raider was that he was a showboating daredevil and, worse yet, a Reservist. He had not been to the Army's Professional Military Education schools, and Ike did not want Doolittle on his staff, but General George Marshall insisted. Doolittle won over Eisenhower. He flew combat missions—leading from the front—but that had to stop when he was briefed on Ultra, the top-secret code-breaking operation revealing German plans in real time. Doolittle proved himself as an organizer and leader, and Eisenhower wanted him on the team for the European Campaign. Doolittle replaced Lt. Gen. Ira Eaker, a Bomber Mafia member, who went to the Mediterranean. One of Doolittle's first acts as commander was to have a sign taken down. It read, "The first duty of the Eighth Air Force fighters is to bring the bombers home alive." He had a new sign made to replace it: "The first duty of the Eighth

---

[7] Civil War heroes have been awarded the Medal of Honor as recently as 2024.

Air Force fighters is to destroy German fighters!" This change in attitude and strategy meant that the AAF would kill the Luftwaffe wherever it was found. This aggressive spirit turned fighter pilots into hunters.

One of Doolittle's first offensives was Operation Argument, called "Big Week," a sustained series of missions against the German aircraft industry. With the Eighth, Fifteenth and RAF bombers striking in coordination, these raids forced the Luftwaffe to make an all-out defensive effort. The Eighth mounted its first thousand-bomber mission on February 20, 1944, as the US kicked off Big Week. It would take more than an hour for a thousand-plane formation to pass over a given point on the ground. Through February 25, 1944, Airmen struck Leipzig, Rostock, Aschersleben, Halberstadt, Gotha, Posen, Schweinfurt, Regensburg, Stuttgart and Augsburg. They hit aircraft factories and component suppliers. The Luftwaffe had to defend all these targets and lost some 500 aircraft and more than 100 pilots, experience that could not be replaced. For the Americans, more than 4,000 bombers had been sent out; 226 were lost, and 2,600 American Airmen were casualties. With fighter escort, bomber losses were still 6 percent. But now the Mighty Eighth was strong enough to bear such losses and return ferociously.

After Big Week, Americans faced a weakened Luftwaffe. Much of the German war machine was built or shipped through Berlin, Germany's biggest city. The first mission against Berlin was planned for March 4, 1944, but most of the bombers turned back and hit targets near the Rhine due to bad weather. A few B-17s from the 13th Bomb Wing apparently did not get the recall order and boldly stormed over the Nazi capital. Fortunately, two groups of P-51s showed up to protect the little bomber fleet. Four of the 32 bombers were shot down, but the psychological victory of hitting Hitler's capital was powerful. Two days later, on March 6, the Eighth Air Force launched another attack on Berlin. About 660 bombers struck their targets. The US lost 69 bombers and 11 fighters, shooting down 81 German fighters. On March 8 and then again on March 9, the Eighth attacked Berlin. More than 100 bombers were lost on these missions, with 1,000 Airmen lost. But as Hermann Goering said after the war, "When I saw those Mustangs over Berlin, I knew that the war was lost."

How accurate was "relatively accurate" bombing in World War II? The average circular error (distance from the target) in 1943 was 1,200 feet. Most bombs fell more than 1,000 feet from their target. Col. Curtis LeMay, commanding the 305th Bomb Group, studied the problem. First, he realized that evasive action did not help bombers avoid flak, but it kept the formation in the line of fire longer and made it far more difficult for bombardiers to see their targets. It was better to fly straight over the target. He identified the best navigators and bombardiers and assigned them to group lead crews. He developed the "combat box" formation to maximize defensive firepower. These became Eighth Air Force standards, and bombing accuracy improved. Eighth Air Force bombers in early 1944 began using a radar system called

H2X, nicknamed "Mickey." Pathfinder aircraft mounted this equipment in a "radome" radar pod under the chin turret or replacing the ball turret. The operator had a radar scope, and when he saw the formation at the bomb release point, the Pathfinder would signal the entire formation to drop its bombs. Although Mickey was not as accurate as visual bombing, it permitted relatively accurate bombing in all weather. Radar bombing helped find targets through overcast but was hardly precise. By 1945, Eighth Air Force put most of its bombs within 1,000 feet of the aiming point.

There was more to the air war than bombing. From the North Africa campaign onwards, airpower supported the ground advance. Tactical air forces established air superiority, interdicted roads and rail lines, destroyed depots and troop concentrations and provided close air support to Allied ground forces. The invasion of Normandy would have been impossible without the decimation of the Luftwaffe, the isolation of the battlefield by wrecking transportation networks, and the pulverizing effect of air attacks on German forces. While all types of fighters and medium bombers (B-25, B-26) were a part of this team, the P-47, with its heavy armament and capacity to withstand damage, proved a highly effective ground attack plane. American aircraft worked with air liaison officers—pilots equipped with radios to contact aircraft—rolling with the ground forces, finding targets, requesting and directing strikes, and sharing updates on enemy locations.

This team was perfected in General George S. Patton's Third Army and Brigadier General Otto Weyland's XIX Tactical Air Command, which set a model for air-ground teamwork that continues today. Weyland retired as a four-star general in 1959, having led forces in Korea and the Tactical Air Command. Elwood "Pete" Quesada flew aboard the *Question Mark*, but unlike his cohorts, he was interested in close air support, using airpower to support engaged ground forces. He led tactical air forces in Italy, then the IX Tactical Air Command in France. In 1946, Quesada was placed in command of Tactical Air Command (TAC), but under the weight of the SAC behemoth, which took the lion's share of money, strength and influence, Quesada retired in 1951. He became the first Federal Aviation Administration chief and was successful in many business endeavors. He died in 1993.

AIRPOWER AT NORMANDY

D-Day would have been impossible without airpower. On the night of June 5-6, 1944, more than 17,000 paratroopers and glider troops went up in (or were towed behind) 1,400 C-46, C-47 and C-53 cargo planes to strike behind the walls of Fortress Europe. The campaign conducted throughout the spring ensured that German troops moved cautiously, usually at night, and always with a fearful eye on the sky. General Eisenhower promised troops that if they saw airplanes overhead, they would be friendly. On D-Day, the Allies flew

15,000 sorties. Eisenhower later said the invasion was based on "the power of the air forces in overwhelming numbers to intervene in the land battle, making it possible for a small force of land troops to invade a continent."

One of the new features of the war was napalm. It was first used in Normandy, as P-38s dropped the jellied gasoline bombs on a fuel depot in Coutances, France, on July 17, 1944. This fearsome weapon would be used heavily in the Pacific, Korea and Vietnam.

Another new aspect of air war was the use of Aeromedical Evacuation. Transports moving wounded out of the battle zones, with nurses and doctors caring for the injured, sick and wounded, was a lifesaving improvement. Many nurses were decorated for valor in these missions. All over the globe, this saved lives. First Lieutenant Mary Louise Hawkins was evacuating patients when her C-47 made a forced landing. A propeller broke loose and severed a patient's trachea. Hawkins made a suction tube from the inflation tube of a life vest and kept the patient alive until help arrived several hours later. All her patients survived. Hawkins received the Distinguished Flying Cross.

More than 1 million patients were evacuated by air in World War II. Only 46 died en route. Seventeen flight nurses gave their lives in the war.

In late July 1944, as the Allies struggled to break out from Normandy, they decided to carpet bomb the enemy. This meant laying down a crushing bombardment to open a path for the Allied advance. Operation Cobra used 1,500 bombers to destroy German fortifications, kill their troops and destroy enemy resistance. It was intended to enable Allied soldiers to unleash their mechanized and armored strength. This worked, but inaccurate bombs killed 111 American soldiers and wounded 500 more. Lt. Gen. Lesley J. McNair, commander of Army Ground Forces, was among the dead.

From Italy, the Fifteenth Air Force and tactical air forces carried out a brutal war, supporting the hard-fought campaign up the Italian boot and hitting France, the Balkans and southern Germany. This extended the range of airpower throughout Hitler's empire. Operation Strangle was a series of interdiction operations during the Italian campaign, aimed at cutting German supply lines north of Rome in early 1944. This operation saw the first use of a radio-guided bomb, the VB-1 Azon, to destroy a bridge, severing a major artery between Germany and Italy.

The Allies drove across northwest Europe, and Operation Market Garden was intended to accelerate the advance with a massive airborne attack over the lower Rhine in September 1944. More than 1,500 Allied aircraft and 500 gliders hauled airborne troops to the Netherlands, but they could not secure bridges and cross the Rhine at Arnhem.

Another brainstorm was Operation Frantic. Bombers would fly from England and Italy, strike Axis targets, then recover in the Soviet Union. The bombers would have some respite, as German air defenses were focused on the approaches from Italy and England. Flying over Eastern Europe was expected to be less costly. Days later, the second half of the mission would see the bombers fly out of Russian airfields, strike again and return to their home bases. On June 2, 1944, a mission took off from Italy, bombed railyards in Hungary, then landed at Poltava, a Soviet airfield in Ukraine. A few of these came off well. Unfortunately, the shuttle missions ended abruptly. Soviet wariness toward Americans devolved into outright hostility when Americans tried to mingle with their Soviet counterparts (the interpersonal interactions were usually amicable); and the obdurate Soviet bureaucracy hampered logistics. Then, the Soviets refused to allow the Americans to support the Warsaw Uprising in August 1944.

Next, a German bombing raid destroyed 47 B-17s on the ground at Poltava, Ukraine. To top it off, Soviet anti-aircraft artillery (AAA) and aircraft attacking American aircraft became a serious problem. Over Yugoslavia, Americans fought Russians when P-38s mistakenly strafed a Russian tank column and, rather than contact the Americans, the Russians attacked them. Both sides lost men and aircraft. The problem was so severe that when President Roosevelt flew to the Yalta Conference in February 1945, the Americans insisted on placing observers at nearby anti-aircraft sites. Operation Frantic foreshadowed the Cold War tensions that would mark the relationship between the USA and the USSR.

The United States and the Soviet Union fought one pitched battle—a dogfight during WWII over the Serbian town of Niš. The outcome wasn't clear and both governments classified details of the incident.

THE DAWN OF ROCKETS AND JETS

Meanwhile, German technology created the V-1 rocket, a cruise missile, and the V-2, a ballistic missile that soared into space before plunging back to earth at more than 1,500 miles per hour. In late 1943, the US identified weapons storage and launch sites being constructed in France's Pas de Calais area and began targeting them. The V-1 began to strike England in June 1944. Some were brought down by anti-aircraft artillery (AAA), but fighters could often intercept them. The V-2 presented a different problem. Other than destroying their manufacturing and launch sites, there was no defense against them. Radio-controlled bombers were tried as flying bombs, to hit the German rocket sites. This failed, and several pilots were killed, as planes blew up prematurely or the radio control apparatus failed. The dead included Navy Lt. Joseph P. Kennedy Jr., brother of future President John Kennedy. Conventional bombing of the launch sites was a better solution. The first V-2s hit London and Paris on September 8. A weapon striking from the sky like a thunderbolt was terrifying, although its military value was debatable.

Thousands of civilians in Britain, France, Belgium and the Netherlands died in these attacks.

Wernher von Braun led the V-2 program. He and many of his team were brought to the United States after the war through Operation Paperclip and worked in the space program. In the 1960s, when von Braun was in charge of NASA's effort to put men on the moon, comedians cracked that he aimed for the stars, but sometimes he hit London.

One of the most impressive German technological advances was in jet aircraft. The ME-262 entered combat in July 1944 and hit hard, downing more than 500 American planes. The AAF found the best way to defeat the jets was to destroy them on the ground. More than 1,400 Me 262s were built, but only a few hundred got into combat. Many could not fly due to a lack of fuel, parts or trained pilots. The first American pilots to shoot down a jet were Maj. Joseph Myers and 2d Lt. Manford Croy, flying P-47s, who forced down an Me-262 on August 28, 1944. Myers went on to become a one-star general, retiring in 1970. Croy was killed in action in April 1945. Fate is fickle.

Germany was testing the ME-262 jet in the summer of 1942. The US got its first jet, the **P-59 AIRACOMET**, into the sky on October 2, 1942, at Muroc Field, California; Robert Stanley was the pilot. The P-59 didn't make it into the war. Its performance was similar to that of the P-51, and it didn't have nearly the range of the Mustang. The first jet-equipped unit, flying P-59s, was the 412th Fighter Group at March Field, California, in December 1945. The P-59 was soon superseded by the P-80, later called the F-80. Test pilot Milo Burcham flew the XP-80 at over 500 miles per hour in level flight on January 8, 1944.

P-51 Mustang

The T-33, derived from the F-80, became one of the longest serving USAF aircraft in history, serving for 50 years with the USAF, flying until 1997.

As the war reached a crescendo—and from this side of history, it's easy to forget that nobody knew how long it would go on or what Hitler might do next—the Eighth launched raids on Berlin (February 3) and Dresden (February 14-15) that killed thousands of civilians, the sort of area bombardments used throughout the war by RAF Bomber Command. Ordered to bomb the center of Berlin, Jimmy Doolittle pointed out that German troops were shifting eastward to face the Soviets, and railyards would be a better target. Eisenhower and Spaatz overruled him: The city center was the target. Thousands of people died. One good thing that came of it was that a Nazi court was bombed. A few prisoners escaped, and the trials of others were delayed. The personification of "Nazi justice," the odious Roland Freisler, was killed. The Dresden bombing was coupled with the RAF incendiary attack on February 13-14. US bombers hit the city on consecutive days, stoking fires that the RAF had started. The death toll reached 25,000. While these raids remain controversial, Dresden was a legitimate target with 100 war production plants and extensive rail facilities. The often-quoted figure of 250,000 dead in that attack originated in the propaganda office of Joseph Goebbels (simply adding a zero onto the actual figure of 25,000 dead). The Soviet and East German governments cynically continued to use this Nazi exaggeration. Such bombings were horrible, even if justified by wartime exigencies. War is often a choice between a bad thing and a worse thing.

Operation Clarion, February 22-23, 1945, was an all-out attack on hundreds of towns, rail facilities, transportation nodes (bridges, road junctions, supply depots) and many lesser targets. The RAF hit the Ruhr, the 9th and 12th Air Forces struck western and northwestern Germany, the Fifteenth Air Force hit southern Germany, and the Eighth Air Force bombed towns in the middle of Germany. More than 7,000 Allied aircraft, from four-engine bombers to single-seat fighters, roared across Germany with impunity. Only a few Luftwaffe fighters challenged the attacks. Thousands of rail cars and motor vehicles were destroyed, hundreds of bridges were dropped and rail lines were cut. Clarion shocked ordinary Germans; bombing accuracy was excellent, damage was extensive and the Luftwaffe was virtually nonexistent.

Operation Varsity was the leap across the Rhine. On March 24, 1945, the Eighth and Ninth Air Forces flew 7,000 sorties supporting the attack. Some 2,000 transport aircraft and gliders delivered troops in the last major air assault of the European war.

The Eighth Air Force ended its bombardment campaign on April 25, 1945, with a strike on the Skoda Works at Pilsen, Czechoslovakia. There were simply no targets left. The Red Army linked up with the Americans, and much of Germany was occupied by the Allies. The last week of the war saw RAF and US bombers dropping food (Operation Chowhound) to still-occupied Dutch people while German commanders agreed not to fire on the bombers. Days later, General Eisenhower accepted the German surrender and sent a laconic message: "The mission of this Allied Force was fulfilled at 0241, local time, May 7th, 1945."

WORLD WAR II: PACIFIC CAMPAIGNS

"If you are going to use military force, then you ought to use overwhelming military force. Use too much and deliberately use too much; you'll save lives, not only your own, but the enemy's too." - Curtis LeMay

War came to Asia in 1937, as China fell under Japan's attack. Claire L. Chennault retired from the Air Corps in 1937, a deaf and disgruntled major. His opinions on pursuit (fighter) use were anathema in the era of bomber ascendancy. He was offered a job in China as it confronted Japan. The US sold China 100 fighter aircraft and Chennault flew the Curtiss P-40 fighter in combat and learned how to use it against the Japanese. His P-40s were decorated with shark teeth and eyes on their aircraft noses.

In 1941, American military pilots were permitted to fly in China. Many who signed up for this American Volunteer Group were, like Chennault, aggressive pilots who did not fit the military mold. Offered generous pay and the opportunity to fight, they became the Flying Tigers. They included Pappy Boyington, Tex Hill and Robert Scott. Entering combat on December 20, 1941, they shot down 300 Japanese aircraft, losing just 12 of their own. In July 1942, the Flying Tigers entered the US military as the 23rd Fighter Group. Chennault was promoted to major general and placed in command of the 14th Air Force, responsible for operations throughout China. Like the Pony Express, the Flying Tigers operated only briefly but left a lasting mark on history.

Air Commandos originated in the China-Burma-India (CBI) theater. Col. Philip Cochran—the inspiration for a character in the *Terry and the Pirates* comic strip—led the first air commando missions against the Japanese in February 1944, supporting troops in Japanese-controlled Burma. With fighters providing close air support and C-47s evacuating the sick and wounded, morale among the ground forces soared. Americans first used helicopters in combat in the CBI. Igor Sikorsky developed the AAF's first helicopter, the XR-4 Hoverfly, in 1942. The helicopter was first used in combat when Lt. Carter Harman of the 1st Air Commando Group flew an R-4 Hoverfly on a mission in April 1944, flying into the Burmese jungle to retrieve four wounded soldiers. There was no other way to get them; the little helicopter could only carry one man at a time. It took four trips over two days to move the men, one by one, but they were rescued, and Harman received the Distinguished Flying Cross. After the war, he returned to his career as a music critic and record producer.

Today's Air Commandos are the Special Operations units. They carry this proud heritage on their missions, often working stealthily behind enemy lines.

"THE HUMP" was World War II's largest airlift. More than 650,000 tons of Allied supplies (more than half was fuel and oil) were delivered from India to China after the Japanese cut the Burma Road. It began on April 8, 1942, as two DC-3 aircraft airlifted gasoline and oil over the Himalayas from India to Yunnan-yi in southern China to support the expected arrival of the Doolittle Raiders' B-25 bombers. The airlift continued till the end of the war. More than 600 aircraft were lost in the operation, an average of nearly one loss every two days. Some 1,659 American Airmen were listed as dead and missing. About 1,200 men were rescued or otherwise got back to friendly lines.

The Curtiss C-46 was the backbone of the Hump. It carried more cargo, had greater range, and handled high-altitude flight better than the C-47. However, it required extensive maintenance, and many were lost in mishaps. C-46s flew in Europe and during the Korean War. The twin-engine Douglas DC-3 was one of the most successful prewar airliners; it would fly in the military as the C-47. The Air Corps ordered its first C-47s in 1940; nearly 10,000 would be in service by the end of the war. They carried cargo and passengers, towed gliders, dropped paratroops and evacuated medical patients. The aircraft carried 28 passengers, about 20 paratroopers, 6,000 pounds of cargo, or 18 stretchers and three medical personnel. C-47s remained in Air Force service until 1975, through the Berlin Airlift, the Korean War and in Vietnam, where they became gunships and psychological warfare platforms.

The first heroes of World War II were in the Pacific, where battered and beleaguered Americans fought valiantly to stem the Japanese onslaught. Capt. Colin Kelly was a B-17 pilot stationed in the Philippines in 1941. During the first week of the war, his crew was credited with sinking a Japanese battleship. Actually, they damaged a cruiser. Under fierce attack, Kelly ordered his crew to bail out while he remained at the bomber's controls. He was killed when the plane exploded in midair. Kelly was awarded the Distinguished Service Cross, becoming a household name in those dark days. Sgt. Meyer Levin, Capt. Kelly's bombardier, was awarded a Silver Star. He survived that fight and was later credited with sinking a Japanese transport ship. He was killed on January 7, 1943 on a mission near New Guinea.

The Swoose is the oldest B-17 in existence, a D model. It saw combat in 1941-42 in the Philippines and the Southwest Pacific, before being converted into a transport for the brass. Capt. Frank Kurtz, her pilot, gave the plane its nickname after many repairs left it a patchwork, taking the name from a popular song from 1941, *"Alexander the Swoose,"* about a half-swan, half-goose bird. Kurtz was an interesting character—an Olympic bronze medalist in diving and a pilot who earned three Silver Stars and three DFCs. *The Swoose* is in the restoration hangar at the Air Force Museum today.

Lt. Boyd "Buzz" Wagner was the first American ace of World War II; he shot down five Japanese planes in the Philippines before being wounded and evacuated to Australia in January 1942. He returned to the United States to train fighter pilots but was killed in an accident on November 29, 1942.

The fall of the Philippines brought one of the most horrific chapters in American military history: The Bataan Death March that followed the surrender of Bataan in April 1942. Thousands of prisoners of war (POWs) were murdered or died of deliberate neglect and brutality. Capt. William Dyess escaped from a POW camp in the Philippines, reached guerrilla forces, and was evacuated by submarine in July 1943. Dyess told the shocking story in a debrief upon his return to the United States. He was training to return to combat when he was killed in a P-38 accident on December 22, 1943. The story of Japanese atrocities was revealed to the American people in January 1944. This hardened American resolve and further embittered Americans toward Japan.

THE DOOLITTLE RAID

Jimmy Doolittle's Tokyo Raid remains one of the most famous feats of World War II. Including Doolittle, just six aircraft commanders on the Tokyo raid flew in the Air Corps before 1941. Fifteen of the copilots were less than a year out of flight school. None of the navigators had two years of experience. On April 18, 1942, Lt. Col. Doolittle led 16 B–25s on a bombing raid against Japan. Each plane carried four 500-pound bombs. The 16 tons of bombs did minor damage, but the mission shocked the Japanese people and their leaders. The raid's most practical effect was that air defense forces were withdrawn to Japan, reducing strength in forward areas.

General Jimmy Doolittle

Most of the Doolittle Raiders survived the mission. Three were killed in action, none due to enemy fire: one died in a bailout accident, two drowned after ditching off the China coast. Eight were taken prisoner (four of these died at the hands of the Japanese). Thirteen more died in combat during the war; four became POWs in Germany. About 50 Japanese civilians were killed in the attack. The Japanese inflicted horrifying retribution against the Chinese, murdering over 250,000 people. Doolittle's planes, supposed to go to American forces in China, were all lost because they had launched much further from Japan than planned. Some crews bailed out, and a few crash-landed or ditched in China. One flew to Vladivostok in hopes that America's Soviet ally would assist them. That crew spent more than a year as virtual prisoners in the USSR before escaping.

The Japanese fought hard. The campaign to capture New Guinea and the surrounding islands took more than two years. On March 1, 1943, US planes spotted 16 Japanese ships in the Bismarck Sea, en route to New Guinea. The Japanese vessels carried 7,000 troops, aircraft fuel and supplies. Over 100 US bombers struck the convoy beginning on March 2. More than 3,000 Japanese troops and sailors drowned, and the fuel and supplies went to the bottom of the sea. In three days, US forces sank eight troop transports and four destroyers.

Paul "Pappy" Gunn spent most of his military career as an enlisted naval aviator. A sailor in World War I, he earned "Wings of Gold" in 1923. He flew in the Navy until his retirement in 1939. He then flew as a civilian airline and cargo pilot in the Philippines. After the Japanese attack, he evacuated US military personnel, was commissioned in the AAF, and was awarded the Distinguished Flying Cross in 1942 for flying medical supplies to the desperate defenders of Bataan. Moving to Australia, Gunn converted A-20 light bombers to strafers by adding four .50-caliber machine guns to their noses, using weapons from wrecked fighters. Gen. George Kenney was impressed and brought Gunn onto his staff. The A-20 ground attack aircraft was so successful against Japanese shipping and ground targets that Kenney had him do the same with a squadron of B-25s. These powerful ground attackers were key in the Solomons, New Guinea and the Battle of the Bismarck Sea, and then hit the enemy in the Philippines and Okinawa. Gunn was highly decorated, retired again after the war, this time as a colonel, and rebuilt his flying business in the Philippines. Gunn died in an airplane crash in bad weather in 1957.

Capt. Jay Zeamer led the most decorated crew. He was the pilot of a B-17 on a reconnaissance mission from New Guinea to Bougainville, the next move in the Southwest Pacific island-hopping campaign of 1943. More than 20 Japanese fighters jumped Zeamer's plane. In a running battle that lasted 45 minutes, Zeamer was severely hit, bleeding from several wounds, and his leg was broken. Six of the nine men aboard were wounded as they fought off attacking fighters. Eventually, the Japanese had to turn back as their fuel ran low. Zeamer nursed the battered B-17 home, landing on an emergency airstrip on New Guinea. He and bombardier Lt. Joe Sarnoski were awarded the Medal of Honor. Sarnoski had enlisted in the Army Air Corps in 1936 and became a bombardier. Severely wounded, he continued to operate his machine gun, even after a 20-millimeter shell knocked him out of his compartment. He struggled back and continued firing at the enemy until he died. He is buried in the National Cemetery on Oahu. All the other crewmen received the Distinguished Service Cross.

REVENGE FOR PEARL HARBOR

One of the boldest strokes of the war was the killing of Admiral Isoroku Yamamoto. He had planned the attacks on Pearl Harbor and Midway, but on April 18, 1943, a flight of P-38 fighters intercepted and shot down a Japanese bomber carrying the admiral. Allied codebreakers had figured out Yamamoto's route and schedule. They knew he would travel on a "Betty" twin-engine bomber, with fighter escort. A flight of eighteen P-38s was sent out to intercept the Japanese formation. The mission was a success: First Lt. Rex Barber and Capt. Thomas Lanphier were credited with the kill, though there is still controversy on the specifics.

As the Americans moved across the Pacific, aerial combat was fierce. The all-time leading US aces both fought in this theater. Major Richard "Dick" Bong was a college student from Wisconsin who became a flying cadet in 1941. Exceptionally skilled, he was reprimanded for stunt flying—looping the Golden Gate Bridge and flying so low in San Francisco that he was blowing clothes off clotheslines—and was not permitted to deploy to England with his group. He shipped out to the South Pacific in the fall of 1942. He did well in combat, logging 21 victories by the fall of 1943. He went home on leave, where he met his future bride, Marge Vattendahl. He returned to the South Pacific in February 1944, broke Eddie Rickenbacker's mark of 26 kills, was awarded the Medal of Honor, and returned to the US that fall for a war bond tour. He returned to combat again, running his score up to 40 victories before General George Kenney ordered him home in January 1945. He was a celebrity, and when he married Marge, the wedding was broadcast on the radio. Bong was assigned to test the P-80 jet fighter. Unfortunately, on August 6, 1945, his plane suffered a fuel pump malfunction on takeoff, and Bong attempted a bailout but was too low to deploy his parachute. Major Dick Bong's obituary shared the front page with the news of the first atomic bomb used at Hiroshima. He was 24.

Major Tommy McGuire was the second leading ace in World War II, with 38 victories in a P-38 in the Pacific Theater. He was awarded the Medal of Honor posthumously for a wild series of dogfights resulting in seven victories in two days, December 25-26, 1944. On January 7, 1945, the 24-year-old major was killed in a dogfight over Negros Island in the Philippines. McGuire AFB, in his native New Jersey, was named for him.

One of the most remarkable characters of the Pacific War was Merian C. Cooper. A bomber pilot in World War I, he was shot down and taken prisoner by the Germans in September 1918. After the Armistice, he found Frank Luke's grave and began the paperwork that led to a Medal of Honor for Luke. Cooper worked for the American Relief Administration, providing food to Poland. As a volunteer, he flew for Poland as they fought off Soviet aggression. Again, he was shot down and became a prisoner of the Soviets. Cooper escaped and was decorated by the Polish government. Returning to the United States, he became a reporter and a documentarian, eventually moving to Hollywood. *King Kong* (1933) was his movie: he was both producer and director. He also produced *Flying Down to Rio*, the first Fred Astaire and Ginger Rogers film. He joined the Army Air Forces in World War II, serving in India and China, helping establish the Hump airlift. Cooper became Claire Chennault's chief of staff and then moved to Fifth Air Force in the Southwest Pacific. He wrote the Medal of Honor recommendation for Jay Zeamer. Cooper witnessed the Japanese surrender on the *USS Missouri* and was promoted to brigadier general in the Air Force Reserve. His postwar movie productions included several John Ford films including *The Searchers* and *The Quiet Man*.

A few Americans shot down enemy planes from all three major Axis powers: 1Lt. Lewis Curdes flew P-38s in the Mediterranean and got seven Germans and one Italian before he was downed and held as a POW in Italy. When Mussolini fell in late 1943, and there was confusion among Italian troops, Curdes escaped. He then served in the Philippines and was credited with one Japanese plane and a U.S. C-47. You read that correctly. Curdes was flying cover for a downed American pilot off the coast when he saw a C-47 headed for an airfield he knew was still in Japanese hands. Curdes tried unsuccessfully to contact the C-47 pilot. Now, thinking it was a lost airplane, he shot up an engine and forced the transport to ditch. As the C-47 crew and passengers got into life rafts, Curdes explained the situation by radio, as Japanese small arms fire from shore splashed nearby. Soon, an American PBY arrived to pick them up. His action saved a dozen Americans from falling into enemy hands.

Lt. Col. Carl Payne was credited with kills from four enemy countries: five German, one Italian and a Vichy French fighter plane before he shot down a Japanese aircraft. Then there was Levi Chase, who flew in North Africa, scoring ten victories over Germans and Italians, then went to the CBI and downed two Japanese. Chase flew combat missions in Korea and Vietnam, tallying more than 500 combat sorties in three wars against at least five enemy nations.

In the spring of 1944, 400,000 Chinese laborers constructed five bomber bases and six fighter fields to support B-29 operations. **THE BOEING B-29** was the most expensive weapons program of World War II, even more costly than the Manhattan Project that produced the atomic bomb. An amazing advance from the B-17, it was faster, carried a larger bombload, and had longer-range than earlier bombers, with a pressurized cabin and tricycle landing gear. B-29s delivered the atomic bombs against Hiroshima and Nagasaki and were the first atomic weapons carriers in the SAC inventory. Some B-29s were later

The B-29 was the most advanced airplane of its time

modified as tankers, and others saw action in Korea.

On June 5, 1944, railroad targets in Bangkok were struck by 98 B-29s taking off from India on the first B–29 raid. Ten days later, on June 15, Brig. Gen. LaVerne G. "Blondie" Saunders led the first Stratofortress mission against Japan, hitting a steel plant at Yawata on Japan's southern island, Kyushu, at the maximum range of the Superfort. Only Kyushu was within range from Chengtu, China. Honshu, the main island, was out of reach. As those bombers hit Yawata, Marines invaded Saipan, which led to the movement of the B-29 fleet to Guam, Saipan and Tinian. Tokyo lay 2,000 miles from Chengtu; it was 1,500 miles from Guam. On November 24, 1944, the first B-29s took off from bases in the Marianas to bomb Tokyo.

Blondie Saunders is a forgotten hero. The 1928 West Point graduate was an all-American football player. The black-haired Blondie (ironic nicknames are a thing among Airmen) was at Hickam Field during the Pearl Harbor attack and got a bomber airborne to pursue the Japanese carriers, although he did not find them. He served heroically at Guadalcanal and throughout the Solomons and New Guinea, being wounded twice. He became one of the first B-29 wing commanders and moved to India in that assignment. Unfortunately, he was severely injured in a crash in September 1944. Curtis LeMay personally flew the mission that located the crash site. Saunders's leg was amputated, and he spent two years in hospitals before being medically retired.

G.I.s expected a bitter campaign to defeat Japan. "Golden Gate in '48!" was a slogan among them, expressing their hope that the war would end in 1948. War plans called for invading Kyushu in November 1945 and landing near Tokyo in March 1946. Horrendous casualties were predicted; at no place had Japanese soldiers given up the fight—each battle was fought to near total annihilation. The US wanted victory as soon as possible. Nothing else was acceptable. Americans were tired of the war, shocked by the rising death tolls and still angry about the dastardly nature of the foe. One of the most popular songs of 1944, Red Foley's "Smoke on the Water," spoke of making "a graveyard of Japan." The shortest path to victory beckoned to the political leadership and the populace.

From Washington, General Hap Arnold controlled the Twentieth Air Force, ensuring that he retained control of the B-29s. This was partly from his responsibility for such a national investment, but also to ensure that the potency of strategic bombing was proven. For the first time, airpower meant global power. Arnold named Bomber Mafia theorist Brig. Gen Haywood Hansell as its commander.

The first B-29 raids from the Marianas were unsatisfactory. The most modern aircraft in the world, with bomb loads no one could have foreseen four years previously, were not hurting Japan's war effort. General Hansell's ideas about daylight strategic bombing were failing miserably. A significant factor was the previously unknown jet stream. Flying at about 30,000 feet, the B-29s encountered winds of 100-200 miles per hour, making accurate bombing impossible with the equipment available. Airplanes flying with the jet stream were achieving ground speeds of 500 miles per hour, while those flying against it were hit by headwinds that left them seemingly motionless in the sky. Additionally, Japan's war industry was not based on giant industrial complexes like Germany's. There was a great deal of cottage industry, as piecework and small craftsmen produced needed materials for the big factories. American planners came to believe that the only way to destroy these workshops was to adopt area bombing tactics similar to those of the RAF over Germany.

A NEW BOMBING STRATEGY

In January 1945, Maj. Gen. Curtis E. LeMay was ordered to replace Hansell. Airpower would be forever different: LeMay was hard-nosed, pragmatic and less tied to dogma. A change of tactics was necessary. Considering the nature of the target—and the fact that Japanese cities were known to be vulnerable to fire—LeMay set aside the daylight strategic bombardment theory in favor of the heretic idea of night area firebombing. He ordered the bomber crews to approach Japan at night from under 10,000 feet, and in a stream rather than a formation. This would reduce engine problems during high-altitude

General Curtis LeMay

flight and the stresses caused by formation flight. The crews were terrified. Going where enemy defenses were most capable without the security of the formations taught to them throughout training seemed suicidal.

The B-29s carried incendiary bombs made of steel pipes filled with Napalm, thermite or magnesium. A B-29 could carry about 1,500 of these bomblets, which broke apart close to the ground. This scattered the flammable material across an area of about 100 feet. They were tried against Kobe and Tokyo in February 1945. The results encouraged LeMay to order a full-scale fire attack on Tokyo on the night of March 9-10, 1945. He sent 314 B-29s to attack from an altitude of 5,000 feet. Launching on the afternoon of March 9, the bombers began hitting the target about midnight. It was the most deadly and destructive air raid of all time, including the atomic bombings. Tokyo authorities estimated that 90,000 people were killed, 40,000 were injured and one million were rendered homeless by the attack. Sixteen square miles of Tokyo were obliterated, roughly 25 percent of the city. US losses were 14 B-29s lost and a few more damaged.

LeMay followed up this assault with fire raids on other major cities. On March 11, Nagoya: over 300 B-29s attacked with no losses. March 13, Osaka: 274 bombers, two lost. March 16, Kobe: 331 Superfortresses burned half the city and left half a million people homeless; three bombers did not return. March 18: Nagoya again. LeMay returned to conventional bombing until his incendiary bombs were restocked. Losses were lower on the night missions, and now Iwo Jima was available as an emergency base. Planes and ships were posted along the bombers' flight path to rescue those who went down in the Pacific, giving crews a chance of being picked up.

The destruction accelerated. By the end of May, more than half of Tokyo had been destroyed. More cities came under attack and daylight incendiary raids were launched. By the end of June, Japan's major cities were essentially charred rubble. General Arnold approved LeMay's plan to attack smaller cities with fire and precision strikes. This was radically different from the war plan that had defeated Germany. LeMay's strategy was aimed at the destruction of Japanese society. In late June, the Twentieth Air Force received an advanced radar, permitting night precision attacks. Now, Japanese oil refineries came under the bombs. In 15 missions, six of Japan's nine oil plants were destroyed. Four B-29s were lost in these raids.

Smaller cities were targeted by the end of July, sometimes two in a night. Many of these towns lacked any effective anti-aircraft defense. The bombers began dropping propaganda leaflets in May and stepped this up in the summer. Japanese cities were warned that attacks were coming. The raids reached new levels of destruction: in Toyama, 99.5 percent of buildings were destroyed.

These raids came at a price. More than 400 B-29s were lost, and about half the B-29 crewmen taken prisoner died at the hands of the Japanese. Henry "Red" Erwin was a radio operator aboard a B-29. During a bombing mission to Koriyama, Japan, a phosphorus smoke bomb ignited inside the aircraft. Erwin grabbed the bomb that burned hot enough to melt aluminum and carried it forward, throwing it out of a cockpit window. Erwin suffered terrible burns but saved his plane and crew. He was expected to die from his injuries, and his Medal of Honor recommendation was expedited—a Medal of Honor was taken from a display at Hickam Field for presentation to Erwin. Somehow, he survived and lived to be 80.

B-29s were excellent minelayers, and one wing was dedicated to this mission, placing explosives in the waters surrounding the island nation to deny use of its coastal waters. Harbors were closed and mines sank nearly 300 Japanese merchant ships (nearly 10 percent of their fleet), a significant economic dislocation. Teamed with the American submarine campaign, Japan was being strangled.

The 509th Composite Group was activated under Col. Paul Tibbets on December 17, 1944. This group's tactics were unusual, and its purpose was a mystery, even to most in the unit. A handful of senior leaders were briefed and understood that the 509th would deliver a new weapon: the atomic bomb.

THE FIRST ATOMIC BOMBS

The first atomic bomb was detonated at Trinity Site in New Mexico on July 16, 1945. With this powerful weapon and an enemy who showed little sign of giving up, it was perhaps inevitable that President Truman would approve its use.

On August 6, 1945, Col. Tibbets flew a B-29 named *Enola Gay* (his mother's name) from Tinian to bomb Hiroshima. About 80,000 people died immediately, and perhaps as many more died later as a result of the bomb. The Japanese government was silent. Three days later, Maj. Charles Sweeney took a B-29 called *Bockscar* (the plane was usually assigned to Capt. Fred Bock's crew) to Nagasaki. The Japanese High Command stood ready to continue fighting, but Emperor Hirohito ordered the government to end the war. Even then, many civilian and military leaders wanted to fight on. There was an attempted coup d'état, resulting in several officers being killed and others committing ritual suicide at the Imperial Palace. Hirohito's recorded speech was broadcast on August 15. He avoided the word "surrender" but announced the end of the war. A bloody land campaign had been averted. The famine that would have struck Japan in a continued war was prevented. Japanese students training for suicide attacks against Americans would live. Japan's murderous occupation of much of Asia ended. Still, the role of the atomic bombings in Japan's unconditional surrender and the ethical quandaries presented by such an indiscriminately destructive weapon are debated.

That the planned invasion of the Japanese Home Islands was unnecessary is clear evidence that "airpower had evolved into a force coequal with land and sea power, decisive in its own right and worthy of the faith of its prophets."

On September 2, 1945, the official surrender ceremony was conducted on the US battleship *Missouri*. Representing the Army Air Forces were Gen. Carl Spaatz, Lt. Gen. George Kenney, Lt. Gen. Jimmy Doolittle, Maj. Gen. Curtis LeMay and a flyover that included 462 B-29s.

It would be several weeks before all of Japan's far-flung forces and garrisons were notified; many refused to believe it. Americans skirmished with Japanese in the Philippines, Guam, and Peleliu in 1946 and 1947. A few Japanese soldiers held out at Guam, New Guinea and the Philippines until the 1970s.

The youngest arm of the military was led by some young men. Many people, then and now, noted how young some Airmen were in World War II. Curtis LeMay was a major general at 38, as was Lauris Norstad. Richard Sanders went from first lieutenant to brigadier general in less than three years. He was 28 when he pinned on a star in June 1944. He spent much of the war flying B-24s in North Africa and Italy, rising to group command and leadership roles in Ninth Air Force headquarters. He retired in 1950. Clinton "Casey" Vincent was 29 when he pinned on a brigadier general's star in 1944. He was a fighter ace and worked for Claire Chennault in China. He died of a heart attack at 40 in 1955.

The colonels were sometimes kids, too. Chesley Peterson was only 23 when he pinned on his colonel's eagles. One of the most dramatic rises was that of John Landers. A cadet when Pearl Harbor was bombed, he was commissioned on December 12, 1941. By 1945, he was a 24-year-old colonel commanding the 357th Fighter Group as a highly decorated ace. Don Blakeslee led the 4th Fighter Group as a 26-year-old colonel. Gerald R. Johnson, a 24-year-old colonel in the Pacific Theater, was an ace with 22 victories when he died in a B-25 lost in a typhoon. He gave up his parachute to save a passenger, who bailed out. There were many entire bomber crews too young to vote. Teenage pilots were common. Lt. Marty Sidener was the youngest AAF pilot in World War II. He completed flight training at 18 in 1944. By the end of the war, he flew 48 missions in the B-26 over Italy and Germany and earned a DFC.

One of the most interesting heroes of the war was Ben Kuroki. Like many young Americans, Kuroki rushed to a recruiting station after Pearl Harbor. A Nebraskan from a farm family, he joined the AAF, but he was initially allowed only to be a clerk because his parents were Japanese immigrants.

In England, however, the desperate need for gunners led to his approval for combat duty. He flew 30 missions, including the August 1, 1943 Ploesti raid. After his combat tour, he visited Japanese internment camps, speaking for the war effort. He then flew 28 bombing missions against Japan, earning his third DFC.

WORLD WAR II FACTS

PERSONNEL

1939: The US had 335,000 men in uniform, just 26,000 in the Air Corps.

1945: There were 12.2 million Americans in uniform. The AAF peaked at 2.4 million in March 1944. About 360,000 of these were in Aircraft Maintenance. That World War II air force was led by 278 generals and 2,550 colonels.

2025: The US Air Force has a total of 318,000 active personnel (plus 170,000 ARC members). Today's Air Force has 240 generals and 3,000 colonels.

AIRCRAFT PRODUCTION

1939: The Army Air Corps had about 1,200 combat aircraft.

1945: The AAF ended the war with 64,000 aircraft. The fleet peaked at almost 80,000 in August 1944. America produced 276,000 aircraft during the war, supplying many to allies. The US built more airplanes than all our enemies combined.

2025: The USAF has 5,400 manned and 400 Remotely Piloted Aircraft.

LOSSES

The AAF lost nearly 23,000 aircraft on combat missions (18,418 against Germany and Italy; 4,530 against Japan). The most significant losses in Europe were on the Berlin mission of March 6, 1944: 69 bombers and 11 fighters destroyed. The worst losses in the Pacific—after 1941—were on a Tokyo mission, May 25, 1945: 26 B-29s were lost.

The AAF's flying safety record is shocking to modern aviators. Many fighter planes (A-36, P-39, P-40, P-38) had accident rates exceeding 100 accidents per 100,000 flight hours. The heavy bombers (B-17, B-24, B-29) all averaged more than 30 accidents per 100,000 flight hours. Twenty-thousand aircraft were lost to "non-combat causes" overseas. In the United States, the AAF lost 15,000 dead in some 50,000 accidents during the war. From Pearl Harbor to the Japanese surrender was 1,366 days. That's about 40 accidents and 10 dead Airmen per day. Some units changed aircraft two or three times during World

War II, even in combat. The Tuskegee Airmen (332nd Fighter Group) flew the P-39, P-40 and P-51, all in 1944. Some men's first sortie in a new plane was a combat sortie. Pilot training was hazardous. A B-24 pilot told me, "It wasn't so bad. Only four men were killed in my basic pilot training class." I asked how many were in the class. "Sixty-four." They faced their mission with stoic determination.

By contrast, today's Air Force Class-A mishap rate (fatality, destroyed aircraft, or damage exceeding $2.5 million) for many years has been around 2 per 100,000 flight hours. Since 2020, fewer than 20 Airmen have died in airplanes. There is still risk in aviation, though it is much less.

CASUALTIES

The AAF lost 40,000 Airmen killed in combat, 15,000 in mishaps, and 12,000 Airmen missing and declared dead. A few prisoners of war "liberated" by the Soviets never returned. About 19,000 Airmen were wounded. More than 41,000 Airmen were captured. There were spikes in casualties in December 1941 (Pearl Harbor) and May 1942 (fall of the Philippines). Then, 1944 averaged 5,700 AAF casualties a month. The last year of World War II was horrifying and bloody. Half of the 5,400 Airmen held by the Japanese died in captivity, compared with 2 percent of the 35,000 fliers in German hands. About 4,000 men were interned when they landed in neutral countries, primarily when their battle-damaged bombers could not make it to their home bases and they chose to land in Switzerland or Sweden. Total AAF combat casualties came to 121,867.

There will never be another war like World War II, due to technology and remotely piloted aircraft. But the valor of men battling to defeat horrendous evils is the Air Force's great legacy.

JIM PFAFF

CHAPTER 5
JET AGE: COLD WAR AND KOREAN WAR

"The first time I ever saw a jet, I shot it down."
- Chuck Yeager

As Commanding General of the Army Air Forces, Gen. Carl Spaatz established three commands that would define the Cold War Air Force: Tactical Air Command (TAC), Strategic Air Command (SAC), and Air Defense Command (ADC). Military Airlift Command (MAC) would surpass ADC in importance during the Vietnam War.

President Harry S. Truman signed the National Security Act on July 26, 1947, while flying on the presidential aircraft *Sacred Cow* (a VC-54). The Air Force was born in an airplane, and it became effective on September 18, 1947. Stuart Symington was the first Secretary of the Air Force, and Gen. Spaatz was the first USAF Chief of Staff. Defense Secretary James W. Forrestal ordered Army Air Forces personnel, bases and materiel transferred to the Air Force in late September. Executive Order 9987, the Key West Agreement, allocated air assets and missions among the armed services. The Air Force's roles were air superiority; air defense of the United States; strategic air warfare; interdiction of enemy land power and communications; combat and logistical air support; intelligence; aerial photography; airlift and air transport; support for airborne operations; interdiction of enemy sea power; antisubmarine warfare and shipping protection; and aerial minelaying.

On April 26, 1948, the Air Force became the first service to begin racial integration, as black Airmen were incorporated into Air Force units. Gen. Spaatz emphasized that the Air Force must "eliminate segregation among its personnel by the unrestricted use of Negro personnel in free competition for any duty within the Air Force for which they may qualify." Thus, the Air Force was ahead of President Truman, who issued Executive Order 9981 on July 26, 1948, banning discrimination based on race, color, religion or national origin in the United States Armed Forces.

It is not surprising that two great civil rights leaders of the 1960s wore Air Force blue: Air Force veteran James Meredith became the first black man to enter the University of Mississippi in 1962. He had earned sergeant's stripes and began college work while stationed in Kansas. Despite threats and hostility, he completed his bachelor's degree at Ol' Miss in 1963. Joseph McNeil, one of the Greensboro Four who staged sit-ins at a Woolworth store's lunch counter seeking integration, was an AFROTC cadet at North Carolina A&T University. He was commissioned and flew as a KC-135 tanker navigator during the Vietnam War. After active duty, he became a Reservist and retired as a major general. His civilian career included financial management and leadership posts in the Federal Aviation Administration.

Progress has always been a hallmark of the Air Force. Less than a month after the Air Force became independent, Capt. Chuck Yeager broke the sound barrier. On October 14, 1947, he flew the rocket-powered X-1 through Mach 1. Yeager was a World War II ace with 11.5 victories—including five in one day. When he was shot down, he evaded capture and worked with the French Resistance. He is best known for his work as a test pilot. He returned to combat in Vietnam as a wing commander and retired as a brigadier general and an Air Force legend. Tom Wolfe's book *The Right Stuff* and the subsequent movie made him famous throughout America. Yeager lived to be 97, the very image of a fighter pilot to the end.

THE JET AGE

Jets developed rapidly. By the end of World War II, the P-59 and P-80 were flying, but jet bombers followed soon. On March 17, 1947, the North American XB–45 flew for the first time, its four jet engines providing a total of 20,000 pounds of thrust. (Each of the KC-135R's four CFM-56 engines delivers more thrust than that!) In 1947, SAC got B–50 bombers, which were essentially improved B–29s with more powerful, more reliable engines and in-flight refueling capability. Still, World War II bombers rapidly became obsolete.

Jets brought the need for ejection seats. They were used first in the F-80B and the B-45. The Luftwaffe had created the ejection seat, and the US worked with British engineers from Martin-Baker to develop this technology. In 1946, First Sgt. Lawrence Lambert became the first American to eject from an airplane, a P-61, over Wright-Patterson AFB. He was awarded the DFC. Lambert had served in the Pacific in World War II and had made 58 parachute jumps before this test. He later worked with Dr. J.P. Stapp and retired as a chief warrant officer. Thousands of aviators have since been saved by ejection from disabled aircraft.

The Cold War was a prolonged period of uneasy truce between two powerful nations. The United States and the Soviet Union squared off for more than 40 years, ready to fight at a moment's notice while the world prayed that it would never come to that. There are, and were then, many who felt the US exaggerated the Soviet threat, and held that the US was to blame for this tense situation. After the Soviet Union collapsed, evidence emerged that the Soviet leadership was as aggressive as Westerners had feared. Indeed, the Germans learned that the Soviets had extensive war plans for the invasion of Western Europe and had even made street signs in Russian for several major West German cities.[8]

The Strategic Air Command became America's shield during the Cold War. Bombers like the B-36 and B-52 were lauded as the war-preventing weapons of the nuclear age, and SAC stood on alert, ready to destroy the enemy. But the goal was not to fight but to force the "Commies in the Kremlin" to say, "Not today, comrade. We cannot attack today." With more than a thousand bombers and hundreds of aerial tankers ready to strike the enemy, SAC became the deadly serious face of the Air Force after the gunfighter drama in Korea wound down.

There was actual fighting in Korea and in Vietnam. Korea continually appeared ready to boil over again. In central Europe, NATO and the Warsaw Pact stood ready, like chessmen awaiting the game. In 1952, the Air Force acquired its first high-speed digital computer, a vacuum-tube-based Univac I. Research and development of missiles accelerated. Reconnaissance flights monitored the USSR and its satellites. The Air Force established its northernmost operational base—Thule AB, Greenland—690 miles north of the Arctic Circle, and manned Arctic outposts that would form the DEW (Distant Early Warning) Line. Aerobee rockets carried monkeys into space. Some survived. It was a time of high pressure and remarkable achievement.

HISTORIC AIRLIFT

The Berlin Airlift, "Operation Vittles," kicked off on June 26, 1948. (The name comes from "victuals," meaning food; vittles is a colloquialism for that term.) When Germany surrendered, the victors divided the country and the city of Berlin into four occupation zones. As the Western Allies (US, Britain, France) planned to create a new German currency, Soviet dictator Joseph Stalin shut down roads and rail lines from western Germany to Berlin. This blockade left two million West Berliners without food and coal for heat. Gen. Lucius Clay, US military governor of Germany, asked Lt. Gen. Curtis LeMay if he could supply the city by air. LeMay retorted, "The Air Force can haul anything!"

The airlift began on June 26, 1948. C–47s and larger C-54s hauled food and fuel to Berlin using established air corridors. Maj. Gen. William H. Tunner, who had led the "Hump" airlift from India to China during World War II, was named commander. Soon, planes launched every 90 seconds from West Germany to Berlin, using air corridors as one-way roads to and from Berlin.

The Douglas C-54 Skymaster was the stalwart of the Berlin Airlift. With its

---

[8] *Washington Post*, March 16, 1993, "E. Germans Had Plans For Invasion -- Documents Indicate Soviet Bloc Seriously Considered Assault."

four engines and ten-ton cargo capacity, it was the primary strategic airlifter in World War II and stayed in service with the Air Force until 1973. A VC-54 was the first military aircraft to carry the president when Franklin Roosevelt's plane, the *Sacred Cow*, took him to the Yalta Conference in 1945.

Lt. Gail Halvorsen flew C-54s during the Berlin Airlift. Flocks of destitute children watched the parade of aircraft transiting Tempelhof Airport, and Halvorsen resolved to drop some candy to them on his landing approaches. Soon, other pilots joined in this endeavor, and the "Candy Bombers" caught the imaginations of the American people, kicking off "Operation Little Vittles." Donations poured in. German children received 20 tons of candy by the end of the Airlift. Halvorsen passed away at age 101 in 2022, a beloved figure in the Air Force and in Germany.

European weather meant instrument flight was a requirement. The flow of aircraft could not be interrupted. If a plane was late or couldn't land in Berlin, it had to continue on the route back to its takeoff base. The Soviets harassed the flights, but most came through safely. The mission consisted of over 120,000 sorties; there were 126 accidents. The Air Force lost 31 men in the operation.

The Soviets gave up the blockade on May 12, 1949. The Allies continued the Airlift for a few months to build stockpiles of fuel, food and medicine in Berlin. Operation Vittles transported more than 2.3 million tons of supplies and 200,000 passengers. The people of Berlin were kept free without war, and the United States proved its resolve to contain the Soviets.

The Berlin Airlift was a learning experience for the Air Force in aircrew and aircraft scheduling, cargo loading and air traffic control. It also demonstrated the need for strategic airlift aircraft.

KOREAN WAR

The USAF was as ready—and unready—as any other branch of the military when the Korean War erupted, but the speed of aviation enabled the Air Force to get into the fight quickly. North Korea invaded South Korea on June 25, 1950. Within two days, USAF transports and fighters were supporting humanitarian evacuations. On June 27, an F-82 pilot, First Lt. William G. Hudson, shot down a North Korean Yak-11, the first aerial victory of the war. B-26s based in Japan bombed rail yards and roads in South Korea, attempting to interdict the onrushing North Koreans. Other B-26s struck targets near Pyongyang, the North Korean capital, the first US action against North Korea itself. In July 1950, B-29s began arriving in Japan and Okinawa. The first strategic attack of the Korean War saw nine B-29s bomb an oil refinery at Wonsan and a chemical plant at Hungnam on July 6, while a month later, 98 B-29s dropped 800 tons of bombs against enemy troops concentrating for an attack on Taegu. U.S.-based fighter squadrons moved toward the fight, and Reserve and Air National Guard (ANG) units were mobilized. As units moved from the United States to Germany, planners wondered if Stalin was using Korea as a distraction from a move against Europe.

Maj. Louis J. Sebille crashed his severely damaged F–51 Mustang fighter into an enemy position on August 5, 1950. Sebille's plane was severely hit during a ground attack, and he rolled in on the target and took it out at the cost of his life. Major Sebille was the first Air Force recipient of the Medal of Honor.

In October 1950, the first airborne operation of the Korean War saw more than 100 C–119s and C–47s drop the 187th Airborne Regimental Combat Team and over 600 tons of materiel north of Pyongyang as General MacArthur's forces drove north. The Chinese intervened in November, as 250,000 fresh troops poured across the Yalu River and smashed into US forces in far northern Korea. The Air Force's role shifted from leading the advance to covering the retreat. Thousands of tons of equipment and supplies, including eight bridge spans for the 1st Marine Division, were airlifted or airdropped as the Marines fought back from the Chosin Reservoir. C–47s evacuated almost 5,000 Marines while Combat Cargo Command evacuated sick and wounded from Pyongyang as the UN forces retreated.

FIGHTER COMBAT

Aerial combat was furious. F–80 pilot Lt. Russell Brown downed a North Korean MiG–15 in the first jet-on-jet dogfight. A force of 70 B–29s dropped incendiaries on Sinuiju, North Korea. A World War II invention, a rudimentary "smart bomb," was revived. The WWII version was called AZON (controllable in Azimuth Only), with a limited radio-controlled steering capability. They were notably effective against bridges. In the Korean War, the Air Force used the TARZON (Tall, Range and Azimuth Only), a 21-foot, radio-controlled, 12,000-pound bomb. As 1951 opened, the Tarzon knocked out bridges while UN forces tried to stem the Communist tide. F–86 Sabres were pulled from air-to-air duties to strike ground targets. The lines stabilized near the 38th Parallel in the spring of 1951, and a bitterly fought stalemate set in. Fighter pilots became heroes in the air, as in World War I. Capt. James Jabara, an F–86 pilot, became the world's first jet ace in May 1951.

Each episode of the TV series *M*A*S*H* opened with a helicopter bringing in wounded soldiers in Korea. Helicopters became bigger, more capable, and longer-ranging, and played a lifesaving role during the Korean War. At least 110 Americans were rescued from behind enemy lines, and more than 20,000 casualties were transported by helicopters. This reduced death rates among wounded soldiers to about half of what they were in World War II. In April 1951, Warrant Officer Donald Nichols led a helicopter-borne Special Operations team into North Korea to recover MiG–15 wreckage. The possibilities of the rotary-winged aircraft seemed endless.

On September 22, 1950, Col. David C. Schilling, a World War II ace, completed the first nonstop flight over the Atlantic Ocean by a jet aircraft (an F-84) from Britain to Maine in ten hours. Schilling refueled three times from KB-29s. This demonstrated the possibilities of rapid deployment with aerial refueling as the force extender. In July 1952, 58 F–84s of Col. Schilling's 31st Fighter Escort Wing completed the first mass flight of jet fighters across the Pacific Ocean, supported by air refueling, from Turner AFB, Georgia, to Yokota AB, Japan. Combat Air Refueling showed its potential on July 6, 1951, as a KB-29M tanker conducted the first combat air refueling over enemy territory, refueling four RF–80 Shooting Stars on reconnaissance missions over North Korea. When B-29s proved vulnerable to MiG-15 jet fighters, F-84s based in Japan took on the daylight bombing mission and were refueled to give them the reach for these flights.

"MiG Alley" was in the northwestern portion of North Korea, near the Chinese border, where the Yalu River meets the Yellow Sea. Relatively close to American fighter bases and closer to enemy bases, it was the combat arena for Soviet-built MiG-15s and the US North American F-86 Sabres. From Manchurian bases, the Communists attacked the Americans, who more than occasionally raced into China in hot pursuit. UN forces were not supposed to cross into China, but it happened. As the war dragged on, the only part that got popular attention were air-to-air operations over North Korea. Some US fighter pilots were World War II veterans, like Gabby Gabreski, William Whisner, George Davis and John Meyer. At the same time, many new heroes made names for themselves in the frigid skies. Capt. Joe McConnell shot down his 16th MiG on May 18, 1953, to become the top-scoring US jet ace. McConnell had been a B-24 navigator in World War II, then attended pilot training. He was killed in a flying accident in 1954. Maj. James Jabara, also a veteran of World War II combat, was close behind, with 15 MiGs, and Capt. Manuel Fernandez came in third at 14.5. F-86 pilots claimed 792 MiGs in air-to-air combat for a loss of 78 Sabres. While this estimate may be high, there is

F-86 in Korea

no doubt the F-86, flown by skilled Airmen, dominated the skies over Korea.

Col. Ralph Parr deserves to be more famous. The only Airman to receive the Distinguished Service Cross and the Air Force Cross, he flew in World War II, Korea, and Vietnam. Parr joined the Army in 1942, was commissioned in 1944, and arrived in the Pacific near the end of the war.

Parr flew two tours in Korea. On June 30, 1953, he shot down two MiGs and drove off ten more, earning the Distinguished Service Cross. Capt. Parr scored the last aerial victory of the Korean War by shooting down an Il–2, for his tenth victory. In 1967, he was with the 12th Tactical Fighter Wing in South Vietnam, flying F-4 Phantoms. He received the Air Force Cross during the siege of Khe Sanh in 1968 for destroying several Viet Cong mortar and machine gun positions in one sortie, despite severe damage to his F-4. This impaired the hostile force's capability to attack Khe Sanh and reduced further losses to friendly cargo aircraft and crews.

With all those losses, the North Koreans must've had quite a lot of pilots. Well, sort of. American pilots reported Russian chatter over the radio. This was connected with pilots noticeably more skilled than others. Dogfights with Soviet pilots in North Korean-marked aircraft were not uncommon. Airmen referred to these adversaries using the Japanese word "honcho," meaning boss.

American pilots who were shot down behind enemy lines and became POWs faced Korean, Russian and Chinese interrogators, and many Americans downed in enemy territory never returned to the United States. While many were likely killed in action, the fate of others remains questionable. For decades, rumors persisted of Americans held in the Soviet Union and China, but nothing was proven. In 1992, Boris Yeltsin sent a letter to the US Senate stating that some Americans were held in the Soviet Union. Unfortunately, no details emerged.[9]

Communist prison camps were hellish, with rampant dysentery and tuberculosis leading to high death rates. Brainwashing (forced ideological training) and torture were common. As the war wound down, Operation Little Switch began, returning POWs who needed advanced medical care, and the Air Force airlifted 150 sick and wounded prisoners from Korea to Japan and then on to the United States. Operation Big Switch began in August 1953, as the USAF airlifted more than 800 former prisoners of war from Korea to the United States after the armistice. The last Airmen to be released were the crew of a B-29 that China claimed had been shot down over their airspace on January 13, 1953. Actually, the plane was over North Korea, and the prisoners were taken by truck across the Yalu. They were released on August 13, 1955.

Forces in Korea have maintained an uneasy truce for over 70 years, one that has occasionally erupted into combat. About 90 Americans have died in fighting there since 1953, along with more than 300 South Korean troops.

[9] *Time Magazine*, June 22, 1992, "A Presummit Gesture".

I toured the Korean War Memorial in Seoul with my crew in 1994. We happened upon a group of primary-school students. Their teacher smiled at us and said something to the children, who then rushed over, hugging us and shouting, "Thank you, Americans!" She told us that the Korean people will never forget what Americans did for them.

One of the major aerospace projects of the 1950s was the Century Series fighters that came to dominate the fighter fleet of the late 1950s and 1960s. The term "Century" comes from their numerical designations, which start at 100. These aircraft drew on lessons learned from Korea. Many early jets looked like World War II aircraft. The Century Series looked different. They were powerful and fast, with afterburners. The speed of sound was no longer a limit. Many of them were unforgiving to fly, and as they pushed the performance envelope, they demanded more of pilots.

THE NORTH AMERICAN F-100A SUPER SABRE was a standout in the Vietnam War. The "Hun" (short for "hundred") was used as an escort for bombardment packages. With a rear cockpit and electronics suite, it was a Suppression of Enemy Air Defenses (SEAD) fighter. As the war dragged on, F-100s were used as Fast Forward Air Control (FAC), supporting ground troops in South Vietnam, most famously as Misty FACs between 1967 and 1970. Losses of the F-100 were stunning because the Hun was a major player in operations in North Vietnam; in the South, it flew low-level missions. These made the F-100 vulnerable to ground fire and surface-to-air missiles, and 240 were destroyed in Vietnam.

THE MCDONNELL F-101 VOODOO had control problems: The plane would pitch up and depart controlled flight when the airflow over the tail was interrupted. This left the controls useless, engines stalled and the jet in a spin. This happened dangerously fast. Its primary role was as an interceptor, guarding North American skies. The RF-101 reconnaissance model saw combat in Vietnam. If the mission called for blazing over a freshly pummeled target for bomb damage assessment, then this was the platform. Of course, everyone on the ground was throwing everything they had at the Voodoo as it raced above Mach 1. Some 33 were lost in combat.

THE CONVAIR F-102 DELTA DAGGER was a cool-looking delta-winged jet, but it couldn't make Mach 1 with external fuel tanks, and at high speed and high altitude, the engine could overheat rapidly. They spent much of their operational life up north, on the Pinetree Line (in Canada), in Alaska, at Goose Bay, Labrador, and Greenland. Only a few "Deuces" reached Southeast Asia.

Its replacement was the **CONVAIR F-106 DELTA DART**. The "Six" was faster, more maneuverable and had a well-designed cockpit. The primary flight instruments were eye-level vertical tapes, easier to read than round gauges. There was an animated, television-like tactical situational display that helped with intercepts and navigation. Flight at Mach 2, up to 60,000 feet, was standard stuff. It was ideal for the alert mission, with a reputation for being a challenging aircraft but fun to fly. Col. Jack Broughton told of an alert on a snowy, minus 28-degree night. At 0207, the "go order" came, and by 0213, 25 of 28 were airborne.

There was one serious problem: ejection seats that didn't work. At least 13 pilots died because of this. Eventually, it was fixed. The last twenty years of the F-106 fleet saw a marked improvement in safety. The Six did not go to war, but its pilots say it would have performed well. On August 1, 1988, the New Jersey ANG retired the last F-106 Delta Darts.

THE LOCKHEED F-104 STARFIGHTER might have been the most glamorous of the Century Series fighters —the missile with a man in it. It could make Mach 2 and climb to 90,000 feet. Pilots loved its steep climb capability. Its looks epitomized the Space Age, futuristic and cool. Drawbacks? It didn't carry much gas and burned that quickly, giving it a short range. Ground crews hated the thin, stubby wings—they were knife-sharp and occasionally sliced into an unwary maintainer. The wingspan was only 21 feet, so it wasn't maneuverable; it was much like a rocket. The US flew the F-104 from 1958 to 1969, purchasing 296, but about 50 were lost in accidents. The US chose the F-106 as its primary air defense interceptor. Despite this, the F-104 proved popular overseas. West Germany bought over a thousand of them, flying the F-104 till 1986. About one-third of the German F-104s crashed, earning it the name "Widowmaker." Other countries that chose the Starfighter were Canada, Italy, Belgium, Denmark, Greece, Norway, Spain, Turkey, Japan, Taiwan, Jordan and Pakistan. Some 2,577 Starfighters were built. Italy flew them until 2004.

THE REPUBLIC F-105 THUNDERCHIEF was built to haul tactical nukes in an all-out war. Its forte was high-speed, low-level bomb delivery. When Vietnam kicked off, the "Thud" was available, a capable fighter-bomber. Two Wings operating the F-105 were established in Thailand, at Korat and Takhli. The Thud became a major player in Rolling Thunder. Thuds hauled tons of bombs North, and 395 of the 833 F-105s built were lost in Southeast Asia. Two-seat Thuds performed the Wild Weasel mission.[10] The F-105 retired in 1984, a much-admired warbird.

You may have noticed there has been no mention of an F-103, F-107 or F-108. The F-103 and F-108 never left the drawing boards. Their beautiful delta-winged designs were expensive, and there were other options. The Navy's A-5 Vigilante was a modified version of the XF-108 design. The F-107 made it to the flightline and was tested but never went into production. Today, people notice its unusual dorsal engine air intake, made to allow a bomb to be carried under the belly.

At the same time, air-to-air missiles were another area of modernization.

---

[10] Missiles that tracked and destroyed enemy radar and surface-to-air missile batteries.

The Sidewinder infrared-guided air-to-air missile was introduced in 1953. Guided missiles offered pilots a method other than gunfire for shooting down enemy aircraft, meaning that long-range missile shots might counter close-in dogfights. There have been many developments since, but the AiM-9 Sidewinder is still a trusted weapon.

FAMOUS TEST PILOTS

Throughout the 1950s, the Air Force was often in the news. Many test pilots were famous then. Among them was Mel Apt. A graduate of the University of Kansas, the Air Force Institute of Technology (AFIT) and the Air Force Test Pilot School, he was involved in flight tests throughout his career. He was flying the X-2 on September 27, 1956, when he became the first man to exceed Mach 3, more than 2,000 miles per hour. Unfortunately, the aircraft left controlled flight, rolling and tumbling out of control. Apt attempted to eject, but the ejection capsule malfunctioned and the flight ended in a fatal crash. Capt. Mel Apt was 32 years old.

Iven Kincheloe was a Purdue graduate who saw action in Korea, earning a Silver Star and ace status at age 23. After Korea, he became a test pilot in Century Series fighters and the X-2 rocket-powered aircraft, where he was the first man to fly above 100,000 feet. On September 12, 1956, Kincheloe flew the X-2 to 126,200 feet, the highest anyone had yet flown. He received the Mackay Trophy. Captain Kincheloe died in an F-104 accident in 1958 when his engine failed on climb out. Early F-104s had a downward launching ejection seat, and in the split second Kincheloe took to roll the airplane, the aircraft lost altitude and he was too low for the chute to open. He was 30 years old.

Iven Kincheloe & Mel Apt with the X-2

One test pilot who had been an Airman in World War II was Bob Hoover. This Tennessean entered the AAF in 1942 and flew fighters in Europe till he was shot down in early 1944. After a year as a POW, Hoover escaped and stole a German FW-190, which he flew to liberated Holland. After the war, he entered flight testing and served as Chuck Yeager's chase pilot on his Mach 1 flight. Hoover went on to a long career in civilian test flight and at air shows.

THE X-15 was first flown on June 8, 1959, by civilian test pilot Scott Crossfield. It was taken to altitude by a B-52 and released for an unpowered glide back to earth. The X-15 was designed to reach near-space altitudes at incredible speeds. Flown 199 times in the 1960s, it provided research data that influenced the Space Shuttle's design. The X-15 was a joint project of the Air Force and NASA. The last X-15 flight was on October 24, 1968.

NASA pilots who flew the X-15 included Neil Armstrong and Joe Walker. Armstrong was a Naval Aviator in the Korean War; Walker was an Airman who earned a DFC in P-38s in World War II. He took the X-15 higher than any other pilot, soaring to 354,200 feet above the earth—that's above 100 kilometers—on August 22, 1963. He died while flying an F-104 chase plane that collided with an XB-70 in 1966.

Seven Airmen flew the X-15. Bill Dana and Joe Walker had moved over to NASA, so five active Airmen flew this amazing aircraft:

Michael Adams flew the X-15 seven times, but was killed in 1967. Descending from 266,000 feet, the aircraft entered a spin at Mach 5. It broke up at a speed of Mach 3.93.

Joe Engle flew the X-15 16 times, earning his astronaut wings. Then, NASA selected him to fly the Space Shuttle.

Pete Knight joined the X-15 program in 1965. He earned a DFC for safely landing an X-15 from 173,000 feet at Mach 4 with total electrical and systems failures. He achieved the fastest speed (Mach 6.7) attained in the X-15 and qualified for astronaut wings.

Bob Rushworth was selected for the program in 1958, flew 34 sorties in the X-15, more than any other pilot, and earned astronaut wings. He also received a DFC for bringing back an X-15 after a major malfunction.

Robert White was the first man to break Mach 6 in the X-15, and the first to top 300,000 feet, thus becoming the first Airman to wear astronaut wings.

Some other Airmen made the news in those days:

Capt. Joseph W. Kittinger Jr., established an altitude endurance record for manned lighter-than-air aircraft, in a balloon over Minnesota for six hours and 34 minutes on June 2, 1957. He spent more than two hours above 96,000 feet, proving a human could survive at that altitude, in preparation for manned space flight. On August 16, 1960, Kittinger jumped from a balloon 102,800 feet over New Mexico to test a high-altitude parachute system. It took him 13 minutes and 45 seconds to arrive on the ground. He fell freely at more than 600 miles per hour for 84,700 feet before opening his parachute. At the time, this was the highest jump and longest freefall on record. He went up with an astronomer in a helium balloon with an open gondola in Project Stargazer in December 1962. They took scientific equipment to an altitude of 82,000 feet and spent 18 hours performing astronomical observations. After these adventures, Kittinger returned to flying. He deployed to Vietnam in B-26 attack aircraft and twice in F-4 fighters. In March 1972, he shot down a MiG-21, but he, too, was shot down on a later mission. He spent 11 months as a POW. Kittinger retired from the Air Force as a colonel in 1978.

Col. John Paul Stapp was an AAF flight surgeon in World War II. His initial research focused on high-altitude flight and unpressurized aircraft, preventing decompression sickness (the "bends"). As aircraft achieved greater speeds in the 1950s, Stapp studied the effects of deceleration, both in crashes and in ejections. He tested his theories by riding a rocket-powered sled on a track at speeds over 600 miles per hour. He suffered numerous injuries in these tests, as the sled stopped in a little over one second, generating as many as 46 G's—greater G forces than any human had ever endured. But he proved that humans could survive ejection from aircraft at supersonic speeds. He showed that backward-facing aircraft passenger seating was safer than forward-facing seats. The airlines didn't heed his words, but the Air Force did, and most USAF passenger configurations have rear-facing seats. Stapp also developed the familiar three-point seat belt used in today's cars.

General Hoyt Vandenberg

Hoyt S. Vandenberg was the Chief of Staff of the Air Force from 1948 to 1953. He was from a wealthy family; his uncle was a prominent senator. A West Point graduate, he flew in pursuit units before World War II. His rise during the war was meteoric, from captain in early 1940 to brigadier general by December 1942. After helping plan the Normandy invasion, he commanded Ninth Air Force. After the war, he was director of Central Intelligence and became Air Force Chief of Staff in April 1948. Personable and smart, he advocated for modernization through technology, jets, rockets, missiles and computers. He selected outstanding leaders: Gen. Curtis E. LeMay at Strategic Air Command, Gen. Lauris Norstad in Europe, Lt. Gen. Otto P. Weyland in Korea, and Maj. Gen. William H. Tunner for the Berlin Airlift. His last year as CSAF saw him battle prostate cancer, which took his life in 1954 at age 55.

JET NOISE IS THE SOUND OF FREEDOM

Marilyn Monroe admired Gen. Vandenberg. Once, she was asked which three men she would like to be on a deserted island with. Her answer was her husband, Joe DiMaggio, Albert Einstein and Hoyt Vandenberg. The general was handsome, athletic, charming and happily married. He was often in the news, frequently a cover photo subject, and something of a celebrity. That's an enviable claim: Marilyn Monroe would like to spend time with you. No word on Mrs. Vandenberg's reaction.

STRATEGIC AIR COMMAND (SAC)

SAC met the challenges of the Cold War with reach, speed, and power. SAC peaked in strength in 1962, with 254,000 troops, 28 percent of the Air Force. It reached highs of 1,854 bombers in 1959, and 1,095 tankers in 1961. It had 1,039 missiles on alert in 1981. Its motto, "Peace Is Our Profession," was ridiculed by fighter pilots during the Vietnam War. SAC Airmen were so thoroughly indoctrinated into the SAC way of doing things—by the book and with a rule for nearly every situation—that they were said to be "SACumcized." This could be positive, ensuring that significant responsibilities were handled effectively, or it could be negative, as some people were immobilized, unable to function without minute guidance. Revered or reviled, SAC dominated the Air Force throughout the Cold War.

SAC displayed its reach in March 1949, as the *Lucky Lady II*, a B–50 bomber, made the first nonstop flight around the world. The flight covered 23,000 miles in 94 hours from Carswell AFB, Texas, with air refuelings from KB-29M aircraft over the Azores, Saudi Arabia, the Philippines and Hawaii. Even though these were essentially World War II aircraft, no place on earth was out of SAC's reach. SAC got its first Convair B–36 Peacemaker heavy bomber in June 1948. An enormous airplane—160 feet long, with a wingspan of 230 feet—the B–36 was built to deliver nuclear weapons. In March 1949, the ten-engine B–36 was tested. Four jets and six reciprocating propellers ("four burning and six turning") enabled it to reach distant targets faster. Though it still cruised at just 240 miles per hour and was not air refuelable, it would stay in the inventory until 1960.

The first jet bomber used by SAC was the **BOEING B-47 STRATOJET**, a six-engine jet bomber designed to strike targets within the Soviet Union. It was Boeing's first swept-wing jet, making its first flight on December 17, 1947. The B-47 could get to its targets swiftly, with air refueling, but didn't carry enough weapons. The answer was a large, intercontinental-range jet bomber, the YB-52 prototype. It first flew in 1952. Still, over 2,000 B-47s were built, pairing the medium bomber with the heavies (B-36, B-52). The B-47 was also a reconnaissance, electronic countermeasures and weather platform. It never dropped a bomb in combat but came under fire while flying reconnaissance missions, and the Soviets shot down at least three. B-47 bombers were retired by 1966, though the RB-47 reconnaissance model served until 1969. In its relatively brief service life, about 200 (10 percent) of the B-47s built were destroyed in crashes.

Range, speed and payload were maximized with air refueling. The KB-29 and KB-50 were basically World War II aircraft, too slow to work with jet bombers. **THE BOEING KC-97** had to fly at maximum speed to refuel a B-47 slowed to its minimum speed, so KC-97 air refuelings were often "toboggans," where the airplanes descended while mated to keep up the tanker's speed. The Air Force tried a KB–47 tanker modification to refuel a B–47 bomber, but the KB-47 could not carry enough fuel to refuel a large bomber. Thus, the **BOEING KC-135 STRATOTANKER** came to be. Boeing's answer derived from its Model 367-80, the ancestor of the 707 airliner and the KC-135. Pairing the KC-135 with the B-52 created a rapidly reacting, globally capable, powerful air deterrent. Air-refueled B-52s could strike any target on earth.

Operation Power Flite was the first around-the-world nonstop flight by jet aircraft. In January 1957, Maj. Gen. Archie J. Old led three **BOEING B-52 STRATOFORTRESSES**, covering a 24,325-mile distance in 45 hours and 19 minutes. They took off from Castle AFB, California, on January 16, 1957, and landed at March AFB, California, on January 18, 1957. They completed four in-flight refuelings with KC-97 tankers flying from Canada, Morocco, Saudi Arabia and Guam. This was the first jet global circumnavigation, and the B-52s flew 24,325 miles in 45 hours and 19 minutes, an average speed of 534 mph.

The big bomber, ready to deliver a nuclear strike, was the backbone of SAC's nuclear war plans. In 1955, SAC began testing alert concepts to deter a Soviet bomber attack. SAC wanted some of its bomber and tanker force on alert, ready to respond in minutes. This became more important as Soviet missiles became a threat. The goal was to get the bombers in the air and headed toward the enemy before the missiles arrived. This alert was initiated on October 1, 1957. It ended in 1991, and no nukes were dropped in a war.

However, the B-52 became the big stick in conventional warfare, dropping bombs in Southeast Asia, Iraq and Afghanistan from the 1960s through the second decade of the 21st century. The -135 has proven to be one of the most capable platforms in the Air Force, serving as a tanker, cargo carrier, passenger mover, medevac, airborne command post, reconnaissance platform and weather observer. The Eisenhower administration's investment in these aircraft was brilliant. B-52s and KC-135s are in their eighth decade of service. Upgrades have extended these aircraft's lifespan beyond the designer's imagination. The B-52 is expected to fly until at least the 2050s. Airmen joke that a B-52 will fly over the retirement ceremony for the B-2.

The first KC-135A Stratotanker was first delivered on June 28, 1957. The jet tanker could cruise at the same speed as jet bombers while refueling, drastically reducing the time for in-flight refueling missions. The last KC-135 was delivered on January 12, 1965. In 1984, Boeing delivered the first KC-135R, modified with more powerful engines and other upgrades. This provided a 60 percent increase in power, a 50 percent increase in offload capability, and a 25 percent reduction in fuel burn rate. Over the next twenty years, the R-model became the standard. The KC-135 remained the only aircraft specifically designed for aerial refueling. More than 350 of the venerable Stratotankers and 76 B-52s are still serving. Wouldn't it be nice if all government spending were so effective?

EARLY WARNING SYSTEMS

It is difficult to describe how real the fear of a Soviet nuclear strike was in the early years of the Cold War. The US and Canada prepared for a Soviet attack by establishing warning radar bands across the Great White North. The Distant Early Warning (DEW) Line was established on July 31, 1957. It was one of three chains of radar sites: The Pinetree Line ran near the 50th Parallel, just north of the U.S.-Canada border across the Great Plains to the Pacific; there was a Mid-Canada Line at about 55 degrees north; and the DEW Line lay north of the Arctic Circle. On August 1, 1957, the North American Air Defense (NORAD) Command, a combined United States–Canadian continental air defense, was established. Along the Atlantic and Pacific approaches to the United States, the Air Force placed its first airborne early warning and control squadrons at McClellan AFB, California, and Otis AFB, Massachusetts, flying EC-121 aircraft.

The Mid-Canada Line went out of service in the 1960s, and the Pinetree Line closed with the end of the Cold War. In 1993, the DEW Line was replaced by the North Warning System. As the Soviet bomber threat lessened in the 1970s, airborne early warning flying units disbanded.

Texas Towers were early warning radar sites 50 to 110 miles off Cape Cod, Nantucket and New Jersey, established in 1956. Their radars provided an extra 30-60 minutes notice of approaching Soviet bombers. Based on oil drilling platforms seen near Texas, the radome-capped towers stood about 60 feet above the water. About fifty Airmen and contractors manned each one. It was difficult duty—the towers swayed in rough seas and storms, the electronic equipment on board

BOMARC: a weapon that was never used

set up a continuous vibration, and one of them collapsed in 1961, killing 28 crewmen. The remaining towers were closed in 1963.

BOMARC, a supersonic anti-aircraft missile capable of intercepting bombers, became operational in 1959. It could hit aircraft hundreds of miles away, destroying them with a nuclear warhead. BOMARC resembled a vertically-launched fighter plane. As the Soviets came to rely more on missiles than on bombers, BOMARC was unneeded. By 1972, it was offline and surviving BOMARCs were used as aerial targets.

Snark seems to be an Air Force thing. You can meet a lot of snarky Airmen. A sarcastic, irreverent attitude isn't unusual in a flying squadron. But the real **SNARK** was the only intercontinental surface-to-surface cruise missile ever deployed by the Air Force. Its origins were in Germany. By early 1945, US experts had reverse-engineered the German V-1 missile, and produced more than 1,000 of them, the JB-2 (jet bomb #2; JB-1 wasn't so hot). They planned to use JB-2s in the ground campaign in Japan. This didn't happen, so the JB-2s were used for testing and as aerial targets. The Cold War Matador and Mace Cruise Missiles were developed from this technology. The Snark was operational for only about a year (1960-61). It was unreliable and inaccurate, missing its targets by miles. One lost missile was found hundreds of miles off course in a Brazilian jungle. By 1961, the Snark had already been made obsolete by ICBMs. Only five of them survive in museums. Many Snarks tested over the Atlantic were lost, leading to the comment that the waters there are "Snark-infested." Airmen can be snarky.

Snark: obsolete as soon as it was fielded

JET NOISE IS THE SOUND OF FREEDOM

Intelligence gathering was a vital part of the Cold War. The Soviets trumpeted their strength and innovation. Their claims were credible, and they had technological boosts from captured Nazis and relentless espionage. The USAF responded with aerial reconnaissance, ground monitoring of communications, and constant readiness for war. American leaders in those days remembered when Pearl Harbor was bombed, and never forgot the shock. The Soviets had been stunned by the Nazi blitzkrieg in 1941, and this colored their Cold War decisions as well. The USAF Security Service was formed in 1948 to monitor and collect radio and other electronic signals from the USSR, its satellites and other countries. While the Security Service has been redesignated and rolled into other intelligence organizations, the name lives on in the Security Service Federal Credit Union in San Antonio. After all, what is more important in a financial institution than security and service?

Operation Looking Glass began on February 3, 1961, when SAC EC-135s began to provide a 24-hour-a-day airborne command post to direct operations if an enemy attack wiped out land-based command and control sites. Fear of a Soviet "bolt from the blue" attack led to measures to counter a surprise attack: ground alert (planes and crews instantly ready for a combat mission), dispersal (putting airplanes at more bases, using more runways to launch the fleet faster), and airborne alert (bombers flying with nuclear weapons). Looking Glass was part of this capability, enabling American forces to fight back even after a savage enemy attack. On July 24, 1990, nearly 30 years of continuous airborne command-post operations (with more than 250,000 hours of accident-free flying) ceased when Looking Glass landed for the final time.

JIM PFAFF

CHAPTER 6
VIETNAM

"We are honored to have had the opportunity to serve our country under difficult circumstances. We are profoundly grateful to our Commander-in-Chief and to our nation for this day. God bless America."
- Rear Admiral Jeremiah Denton, returning POW

Few events have done so much harm to the United States as the Vietnam War: Political and social division, lies told by leaders, a war in which military rationale was often lost amid diplomatic gamesmanship and outright treachery, and 58,000 Americans dead for a cause American elected leaders did not even believe in. Yet, Americans fought heroically and those in uniform demonstrated more honor than those who sent them to war.

Lt. Karl Richter

Karl W. Richter was a 1964 Air Force Academy graduate. He went to pilot training, then checked out in the F-105. He excelled in training and arrived at Korat AB, Thailand, in April 1966. He flew his first mission over North Vietnam four days after he arrived. On September 21, 1966, he got behind a MiG-17 and closed to gun range, putting cannon fire into the right wing and tail of the enemy fighter. Just 23, Richter was the youngest American pilot to shoot down a MiG over Vietnam. Richter flew his 100th mission on October 13, 1966, and received approval for an additional 100 combat missions.

Richter's mission on April 20, 1967, was to lead a flight of F-105s against

AAA (anti-aircraft artillery) and SAM (surface-to-air missile) sites protecting an important North Vietnamese railroad facility. Once they had suppressed the defenses, US forces would pound the rail hub. But when Richter and his flight arrived in the designated target area, they found it shrouded by clouds. Richter repeatedly attacked, and under heavy fire his flight destroyed and damaged several SAM sites, enabling the strike flight to destroy the target. He was awarded the Air Force Cross.

Richter flew 198 missions against North Vietnam. But his luck ran out on July 28, 1967, as he took a new pilot up as his wingman in Route Package 1—the "safest" area of North Vietnam. Rolling in on a bridge, his plane was hit by AAA. As they turned for Korat, Richter called "May Day!" and ejected. He disappeared into the clouds. A rescue helicopter was nearby and immediately went after him. When they arrived, Richter was severely injured, battered by the rocky terrain where he landed. He died before he reached a hospital.

When asked why he volunteered for another tour, Richter said, "It will seem strange to say it, but I kinda believe in what we're doing over here. I believe in the work." Karl Richter seems a man from another time, when duty, honor and country were uplifting words and not a punchline. His story should be known.

One of the earliest missions of the Vietnam War was Operation Ranch Hand, an attempt to defoliate much of the Vietnamese jungle. The idea was simple: deny the enemy cover, remove his hiding places, expose his movement and reveal his lines of communication and supply. The Ranch Hand crews flew hard. With two dozen little C-123 cargo planes, they delivered 19 million gallons of herbicides. The mission itself was risky, even without enemy action. An engine failure on takeoff was catastrophic. The spray patterns often required steep turns at low altitude in mountainous terrain.

The spray mission was the idea of the civilian "Whiz Kids" advising Presidents Kennedy and Johnson. They sought elaborate technological solutions to one of humanity's oldest and crudest problems: hostile aggression. Defoliation was one of these brainstorms. In November 1961, the Joint Chiefs expressed concerns about this operation: Was it chemical warfare? Would the local people be harmed? Washington ordered it to be done anyway.

The first mission was flown on January 13, 1962, using C-123 cargo aircraft, a plane that originated as an assault glider after World War II. Given a pair of piston engines, the little craft was a handy machine for short cargo hops, moving troops, and able to work from small, unprepared airfields. Adding two small jet engines on the wings made the little plane even more capable. The C-123 played several roles, but its notoriety comes from its role in Ranch Hand. Cruising at 130 miles per hour at 150 feet above the ground was an invitation to enemy gunners. The airplanes often flew in formation to lay down the herbicide. Weather and temperature affected the efficacy of the chemicals. Calm, cool mornings were ideal for saturating the targeted areas.

On Feb. 2, 1962, the Air Force lost its first aircraft in Vietnam, a Ranch Hand C-123. All three crewmembers died in the incident. But Ranch Hand was supported by Defense Secretary Robert McNamara, and his data indicated that it was achieving its aims. One sign that it was working was increased ground fire. Forward Air Controllers and fighters were assigned to escort the sprayers. Ranch Hand aircraft took a hit every six missions. Five were downed in combat. The Air Force Museum displays *Patches*, an aircraft that took hundreds of hits from enemy fire and whose crewmen earned seven Purple Hearts.

Defoliation became politically unacceptable and ended on January 7, 1971. The health issues arising from the use of these chemicals continue to anger and worry Vietnam veterans and their families. The Vietnamese government estimates that four million people were exposed to these defoliants and that 400,000 have died from their effects. Birth defects, cancers and genetic issues are all linked to Agent Orange.

Tactical airlifters were essential in Southeast Asia. Cargo planes like the C-123, C-130, and C-7 Caribou enabled troops to move quickly and be well supplied. People are familiar with flying reindeer, but flying Caribous were a mainstay in Vietnam. The little twin-engine transports flew from tiny dirt fields, to move troops, bring out casualties and provide supplies to otherwise inaccessible outposts. The **LOCKHEED C-130 HERCULES** entered the Air Force in 1956. This four-engine turboprop airlifter could fly 2,500 miles, carry almost 50,000 pounds of cargo or up to 92 troops, and take off and land on just 3,600 feet. Flying in every kind of weather and terrain, and under enemy fire, 65 "Herks" were lost in Vietnam.

C-130: one of the greatest airplanes in history

HEROIC AIRMEN

Joe Jackson enlisted in March 1941 as a mechanic and became a pilot. He flew F-84s in Korea and the U-2. He was flying a C-123 on May 12, 1968, when a Special Forces camp at Kham Duc was being overrun. An O-2 observation aircraft[11], two helicopters, and an A-1 Skyraider[12] all went down in minutes. A C-130 made it in and tried to depart with 189 souls on board. It was shot down on takeoff, the worst loss of life in any US military aircraft incident. Still, a few C-130s and helicopters bravely continued their missions, and rescued hundreds. Late in the afternoon, it appeared that the Americans and South Vietnamese who were alive had been evacuated. But then came a call: three Americans, the combat control team directing the evacuation, were still on the ground. Into the maelstrom flew Lt. Col. Joe Jackson, diving his C-123 rapidly and landing with enemy gunners shooting at him from close range. Debris littered the runway, leaving only about 2,200 feet for Jackson to use. He hit the brakes and the three Airmen leaped into the open rear cargo door. A 122 mm rocket suddenly rolled across the runway and, incredibly, did not explode. Jackson taxied around it, applied full power, and got back into the air under intense fire, having spent less than one minute on the ground. He received the Medal of Honor.

Another innovative use of a cargo plane was the gunship. On December 15, 1964, an FC-47 (later designated AC-47) with machine guns mounted in its cargo bay flew the first gunship mission in Vietnam. By mid-1965, a squadron of AC-47s, each carrying three 7.62 mm miniguns, was in Vietnam. It was said that a ten-second pass could put a round onto every square foot of a football field. Soon came AC-119 gunships with more firepower, and then the AC-130, with even more guns blazing, including cannons ranging from 20 mm to 40 mm and a 105 mm howitzer. Gunships are still an important weapon in counterinsurgency and antiterrorism.

An AC-47 gunship crew could drop flares to illuminate enemy positions. On February 24, 1969, A1C John L. Levitow was on a gunship hit by enemy fire. One of these flares was dropped by a wounded comrade and was rolling around in the cargo compartment. Levitow threw himself on it, picked it up, and hurled it out the cargo door an instant before it ignited, saving the aircraft and crew. Levitow was awarded the Medal of Honor.

Courage in helping others is a remarkable display of character. On March 10, 1966, Maj. Bernard Fisher and another pilot were providing air support for a Green Beret camp under heavy attack. Maj. Dafford "Jump" Myers was hit and crash-landed on the airstrip. Fisher landed his A-1E Skyraider near the damaged plane. He parked momentarily and, under fire, climbed out of the cockpit to pull Myers in. Then, with his plane taking hits, he took off. Major

---

[11] A Cessna light aircraft with engines inline in the aircraft's nose and behind the cockpit.
[12] The A-1 Skyraider was a propeller-driven plane that carried a potent weapons load to support ground troops.

Fisher was awarded the Medal of Honor.

The Pararescue Jumper (PJ) came into its own in Vietnam. CMSgt. Duane Hackney is the most decorated enlisted Airman in Air Force history. He joined the Air Force in 1965 and went to Vietnam in September 1966 as a PJ. Hackney flew over 200 combat search and rescue missions and received numerous decorations for valor. He received the Air Force Cross for a mission on February 6, 1967. Sent to rescue a pilot shot down over North Vietnam, at first Hackney was unable to locate the pilot. He descended with a litter on a second attempt and retrieved the man. They were back aboard when the HH-3E "Jolly Green Giant" helicopter[13] was hit. Without hesitation, Hackney removed his parachute and put it on the rescued pilot. As he grabbed a spare parachute, the helicopter was hit again and blew up. Hackney was still donning the chute, but survived the bailout and was rescued. His Air Force Cross citation states, "Airman Hackney chose to place his responsibility to the survivor above his own life." Combat Search and Rescue (CSAR) forces saved nearly 4,000 lives at the cost of 71 CSAR men during the Vietnam War. Hackney was "service before self" personified.

CMSgt Duane Hackney who put others first

A1C William H. Pitsenbarger was a pararescue jumper in Vietnam. From August 1965, Pitsenbarger completed more than 250 missions. Once, he hung from a cable suspended from a helicopter to rescue a wounded South Vietnamese soldier from a minefield. On April 11, 1966, a US infantry company was surrounded by a Viet Cong battalion. Getting help through the jungle canopy looked impossible. Air Force HH-43 Huskie helicopters were equipped with cables and winches, enabling them to perform the mission. Pitsenbarger volunteered to be lowered to the ground to help. He got wounded men onto litters where they were hoisted up, saving at least nine soldiers. The battle-damaged helicopter had to leave, but Pitsenbarger refused evacuation, giving his place to a wounded soldier. The battle raged on through the night. Pitsenbarger cared for wounded men, was wounded three times and died fighting alongside those soldiers. In the morning, US forces were able to recover survivors and the fallen. William H. Pitsenbarger received the Air Force Cross, later upgraded to the Medal of Honor.

---

[13] The HH-3E got its nickname because it's a very big helicopter and it's green. In the 1960s, the "Jolly Green Giant" was featured in TV commercials for canned vegetables. Airmen tied the two big green figures together.

A less intense but welcome aid to downed Airmen was the Madden Kit. In February 1971, A-1 Skyraider pilot Capt. James Madden was involved in a multi-day rescue mission over Laos. Returning to cover the downed pilot, he learned that the man's radio batteries were dead. Capt. Madden worked with maintenance and life support to fill an empty flare canister with a survival radio, batteries, water and miniature Snickers bars. These "Madden Kits" were dropped to survivors on many rescue missions.

MISTY FACS flew F-100s fast and low over the enemy, armed with cannons and marking rockets, to direct strike aircraft onto targets. They attacked SAMs, AAA sites, trucks, bridges, boats and anything moving on the Ho Chi Minh Trail. It was dangerous work: of the 157 Mistys, 42 were shot down (28 percent). Eight were killed in Vietnam. Four became POWs. Col. Bud Day received the Medal of Honor. Two became Chief of Staff of the Air Force—Ron Fogleman and Merrill McPeak. Four others wore generals' stars; thirteen became colonels. Two became astronauts, and a couple became physicians. Dick Rutan flew the Voyager aircraft around the world. They are legendary.

Operation Rolling Thunder was a tragedy, a misuse of airpower that cost thousands of lives in a misguided effort to show the North Vietnamese that America was serious about defending South Vietnam. If a nation goes to war, there will be deaths and destruction. This violence has to attain some transcendent goal or at least inflict one's will on the enemy, not merely show "resolve."

A1C William Pitsenbarger gave his life that others might live

JET NOISE IS THE SOUND OF FREEDOM

Rolling Thunder was an aerial bombardment campaign conducted by the United States against North Vietnam from March 1965 until November 1968. The campaign was often paused to give the enemy time to reflect on their suffering or to open negotiations. Naturally, the North Vietnamese used the pauses to repair, resupply and reinforce their air defenses and targeted areas. The Johnson administration famously directed the campaign from the Oval Office, imposing strict limitations on which targets could be attacked, what weapons could be used, and even routing and timing, hoping to increase pressure on Hanoi gradually and "persuade" the Communists to come to terms. Targets had to be approved by the White House and could not be within 20 miles of Hanoi or Haiphong. At times, the MiG bases launching enemy fighters were off limits. Even more frustrating, a target on the approved list one week could be forbidden the following week. Worse yet, Air Force leadership limply accepted these tactical decisions. The value of the objective and its costs, in terms of lives and aircraft, did not seem to matter.

This idea of weakening the enemy's resolve was widely seen as nonsense by Airmen. A sharp, punishing offensive—what would be called shock and awe in a later war—was a better use of airpower, tactically and strategically. Rolling Thunder went against military experience and expertise. Still, McNamara made it clear that the Whiz Kids' quantitative analysis approach to war was how it would be. At the same time, Lyndon Johnson and Robert McNamara decided airpower could not win the war and poured thousands of troops into Vietnam over the next several years. President Johnson restricted the bombing of North Vietnam on March 31, 1968.

American Airmen fought heroically throughout the campaign over North Vietnam. Americans engaged in aerial combat under restrictive rules of engagement, losing 922 aircraft over North Vietnam and Laos. Daring rescues brought some aircrew out of enemy territory, but hundreds died, and hundreds more ended up in squalid POW camps. The fate of many is still unknown. More than a thousand Americans are still listed as Missing in Action. Most likely, they perished in the jungles, but some men's fates remain a question.

Self-imposed restrictions meant that airpower did not pose a threat to the Communist government or its ability to support the war in the South. Gradually escalating the pressure enabled the enemy to adjust and strengthen their defenses. Perhaps a sharp, devastating air offensive in 1964 or 1965 could have forced North Vietnam to real negotiations. Rolling Thunder was never intended to do that. As one studies the Vietnam War, it is little wonder that Americans lost faith in their institutions and the government.

Operation Bolo took place on January 2, 1967. Col. Robin Olds led one of the boldest actions of the Vietnam War. The radio chatter and electronic signatures of a strike force were simulated to lure North Vietnamese MiGs up, where a large fighter formation—F-4s of the 8th Tactical Fighter Wing—

*Brig. Gen. Robin Olds:
A leader all his life*

was waiting for them. Olds' Wolfpack shot down seven MiG-21s that day, the most successful air-to-air combat of the Vietnam War. Robin Olds shot down a MiG-21 to become the only USAF ace with aerial victories in World War II and Vietnam.

Olds was the son of one of airpower's early theorists, Robert Olds. Graduating from West Point in 1943, he flew fighters in Europe, logged 13 aerial victories, and was promoted to major, commanding a fighter squadron in combat at just 22 years old. He moved quickly through peacetime command and staff jobs, rising to colonel at age 31. In 1966, Olds became commander of the 8th Tactical Fighter Wing, fighting in Vietnam. His aggressive command style was electric: morale—and results—soared. His pilots emulated his large, non-regulation mustache. Enemy pilots were scourged by the "Wolfpack," as Olds himself recorded four kills. Returning to the US in the fall of 1967, Olds met with the Chief of Staff of the Air Force and President Lyndon B. Johnson. His advice: "Get us out of this war." Johnson said, "How?" Olds' reply: "Win it!"

Olds, like Karl Richter, represents American Airmen sent to fight in a cause their superiors did not fully commit to winning.

BATTLE OF THE BRIDGES

On April 5, 1965, the Air Force launched a strike against the Thanh Hoa Bridge, crossing the Song Ma River in North Vietnam, 70 miles south of Hanoi. It was on the North Vietnamese supply route to the south; narrow, just wide enough for a rail line, tough to hit with iron bombs. The strike failed: 79 US aircraft were sent, but at the end of the day, the span stood. Two F-105s were shot down. Over time, hundreds of sorties were targeted against the bridge nicknamed the "Dragon's Jaw," but it remained standing. In May 1972, laser-guided bombs were available, and a strike by F-4s rendered the bridge unusable. Navy A-7s finally took it out in October 1972.

Laser-guided bombs destroyed the Paul Doumer Bridge connecting Hanoi to the port city of Haiphong on May 10-11, 1972. Airmen had struck at that viaduct repeatedly in 1967 with little effect. But with "smart" bombs, airpower was able to live up to its promise.

Air refueling was vital to completing missions from Thailand to North

Vietnam. Dozens of air refuelings each day enabled hitting distant targets and returning to base. Tanker crews sometimes ventured into North Vietnam unarmed, to rescue damaged aircraft, or those who had used up their fuel in the fight. On May 31, 1967, Maj. John Casteel and his KC-135 Stratotanker crew were on a routine mission over the Gulf of Tonkin, planning to head to Okinawa after refueling a couple of F-104s. US Navy aircraft, dangerously short of fuel, arrived unexpectedly. The KC-135 refueled two Navy KA-3 tankers, two Navy F-8s, two Navy F-4s, and its assigned F-104s. At one point, the KC-135 was refueling a KA-3 while the KA-3 was refueling an F-8, a three-level air refueling. The tanker crew had saved six Navy aircraft and ten men, and earned the DFC and the MacKay Trophy for the most significant aerial flight of 1967.

Capt. Bob Pardo and Lt. Steve Wayne were in an F-4 Phantom strike flight against a steel mill near Hanoi on March 10, 1967. Their aircraft and another took flak as they approached, but they hit their target and began their egress. The F-4 flown by Capt. Earl Aman and Lt. Bob Houghton was leaking a lot of fuel. Bailout over North Vietnam meant death or capture. If they could reach Laos, a rescue might be coordinated. It wasn't unusual for a crew to stay with a damaged buddy. But Aman's plane flamed out over North Vietnam. Pardo slid aft of Aman's F-4 and had him drop his tailhook. Pardo then put his windscreen against the tailhook, allowing him to keep Aman moving forward with a slower descent rate. Over and over, Pardo had to maneuver back into position to keep pushing, but they were getting closer to help. Then Pardo had an engine fire. Down to 6,000 feet over Laos, with CSAR forces on the way, both crews ejected. All four men were picked up.

F-4 Phantoms – a workhorse fighter

The wingman concept is part of the Air Force ethos. It means taking care of other Airmen, being counted on to support one another. It's hard to think of a better example of the "wingman" concept than "Pardo's Push." But wouldn't you know, someone considered it a waste of an airplane. Pardo was nearly court-martialed for the loss of his F-4, but his wing commander, Robin Olds, convinced the 7th Air Force commander to drop the charge. Years later, Pardo and Wayne were awarded the Silver Star.

The Wild Weasel mission was another heroic task. The Weasel motto, YGBSM—"You Gotta Be Sh!tting Me"—was said to have been muttered by a former B-52 Electronic Warfare Officer (EWO) when he was told what the mission would be: be illuminated by enemy air defenses to find and destroy them, thereby protecting the rest of the strike force. The Weasels provoked enemy anti-aircraft defenses into targeting them, then used those radar signals to find and destroy the enemy, essentially playing chicken with missiles. SA-2 SAMs have a range of 17 miles, while the Shrike missiles the Weasels used against SAM sites had a range of about 10 miles, meaning they had to get well inside SAM range to strike. Attacking from low altitude—below 3,000 feet—put the Weasels below radar coverage but down where enemy AAA filled the sky with shrapnel. Crews devised the "Shrike toss," climbing steeply and firing the missile at the top of the climb, stretching its range. Later, the Standard Anti-Radiation Missile (ARM) came into use, with a 50-mile range.

Weasel crews—a pilot and a navigator trained as an EWO—were highly decorated. Fifteen Weasels received the Air Force Cross, but they lost 48 aircraft, with 34 crewmen killed and 19 taken prisoner. Two Weasel pilots were awarded the Medal of Honor: Capt. Merlyn Dethlefsen earned the Medal of Honor on March 10, 1967. When his flight leader was shot down during an attack on a steel mill north of Hanoi, Dethlefsen took command. Despite battle damage, he and his EWO, Capt. Mike Gilroy, made five passes to suppress SAMs, destroying two SAM sites in a furious ten-minute duel with the missiles and flak. Gilroy received the Air Force Cross.

Maj. Leo K. Thorsness and his EWO, Capt. Harold E. Johnson, were leading a Wild Weasel flight on April 19, 1967, when they were attacked by MiGs, SAMs, and AAA. They shot down one MiG and may have gotten another. After using all their missiles and bombs, they air refueled and returned to cover a rescue, armed with only their 20 mm cannon. Thorsness received the Medal of Honor, and Johnson received the Air Force Cross. They were shot down a few days later and spent six years as POWs.

I don't know the rationale behind two pilots receiving Medals of Honor while the two EWOs sitting a couple of feet behind them, in the same fight and just as critical to mission success, did not receive the same medal.

The Air Force Cross (AFC) is the second-highest valor award, given for extraordinary combat heroism. About 180 were awarded, but only a few recipients earned multiple awards. The only Airman to receive three Air Force Crosses was Col. James H. Kasler. A hero might have flown in the bombing campaign against Japan or been an ace in Korea. A hero may have flown some of the most harrowing missions over North Vietnam or endured the agony of nearly seven years as a prisoner of war (POW). Jim Kasler did all of that. A 19-year-old B-29 tail gunner in 1945, he later became a pilot and flew F-86s, downing six MiGs in Korea. Vietnam was his third war. Six months into his tour, in August 1966, *Time Magazine* called him a "one-man Air Force." He earned his first Air Force Cross destroying an oil facility in Hanoi in his F-105. On a later mission, when a buddy was shot down, Kasler located the man and flew cover for him. Rather than going home when his fuel ran low, Kasler went to a KC-135 tanker, then returned to watch over the downed pilot. Unfortunately, Kasler was then brought down in a hail of cannon fire. Kasler spent 2,400 days in captivity, with minimal medical care for his severe injuries, enduring brutal torture in the lonely hell of solitary confinement. In 1973, Kasler received an AFC for his valiant attempt to save that pilot, then a third for his incredible resistance while a prisoner of war.

Three Airmen received two Air Force Crosses: Capt. Leland Kennedy was the pilot of an HH-3E of the 38th Aerospace Rescue and Recovery Squadron (ARRS) on October 5, 1966, on a rescue mission in North Vietnam. After the lead helicopter was severely damaged, Kennedy assumed command of the rescue. His aircraft was hit, and two crewmen were wounded, but on the fifth attempt, they picked up a survivor. Two weeks later, Kennedy earned a second Air Force Cross. Amid heavy ground fire, he landed and recovered the crew of a downed helicopter and the survivor they had rescued. In the face of continuous enemy attack, Captain Kennedy then recovered a second downed pilot, for a total of six men saved on this flight.

Capt. John Dramesi received two Air Force Crosses. The first came when he was shot down while attacking a SAM site. On the ground, he continued to call in strikes on the enemy, keeping rescue forces from harm. He was taken prisoner. After a second escape attempt, he endured some of the most sustained and horrific torture of any POW in North Vietnam. For 38 days, he was starved, beaten and tortured around the clock. His co-escapee, Maj. Edwin Atterbury, died under this brutality. They received the Air Force Cross for their heroic effort and resistance to the enemy.

Lt. Col. Robinson Risner was the first living Airman to earn an AFC. This was due to his heroic leadership during the first attack on the Thanh Hoa Bridge in 1965. He was featured on the cover of *Time Magazine*. A few months later, he was shot down over North Vietnam, and as a senior ranking officer, he inspired and led the POWs. He was singled out for ruthless treatment but successfully resisted the enemy and set an example for his fellow prisoners, earning him a second Air Force Cross.

The **MCDONNELL-DOUGLAS F-4** was one of the longest-lived USAF fighters, a mainstay of the Vietnam War and still in service through Desert Storm. The Navy began its development in the late 1950s. The Air Force was instructed to accept the F-4 as part of Defense Secretary Robert McNamara's initiative to standardize the military's fighter fleet. In May 1963, the Air Force received its first F-4, which hit Mach 2 on its first flight. Airmen warmed to the hulking fighter and became its primary user. Soon, the new jets were in Southeast Asia, flying air-to-air missions, ground attack in South Vietnam, as bombers in Laos and North Vietnam, as a reconnaissance platform and as Wild Weasels.

The highest total of USAF Phantoms in Southeast Asia at any one time was 353, reached in 1972. Over the course of the war, the USAF lost a total of 445 F-4s: 193 were lost over North Vietnam, most to AAA and SAMs. F-4s claimed 107.5 air-to-air victories in the Vietnam War.

F-4s began leaving active duty as F-15s and F-16s entered service, and by 1987, only RF-4Cs and F-4G Wild Weasels were in the active force. These planes flew in Desert Storm, and the last F-4 sorties came in 1996, when two F-4G's left Boise, Idaho, where they had served with the ANG. Now, there may be someone who remembers, "I saw an F-4 over the Gulf of Mexico in 2010 or so..." Some F-4s were converted into drones at Holloman AFB, New Mexico, and Tyndall AFB, Florida, and used as aerial targets until 2016.

BAT-21

Lt. Col. Gene Hambleton was near retirement when he was assigned to fly as a navigator on EB-66 electronic jamming aircraft. These large, unarmed aircraft jammed enemy air defenses and collected intelligence on them. While escorting B-52s, Hambleton's plane was shot down by an enemy surface-to-air missile. For most Airmen, surviving a shootdown meant hoping for a rescue attempt. For Hambleton, it became an epic effort. He had been an aircraft mechanic before entering navigator training in 1944. He did not see combat in World War II, but he flew 43 combat missions in the Korean War in B-29s. He then flew the B-47 and went into missiles, commanding a Titan II squadron. His background in nuclear missiles and electronic warfare made it imperative not to let him fall into enemy hands. Down in the jungle, he was in an area full of North Vietnamese soldiers moving south in their spring offensive. He evaded capture, following a FAC's instructions, which were often described as golf holes: "Bat-21B, do you know the first hole at Tucson? Move like that." Hambleton was an avid golfer and would know to walk straight for 300 yards, then dogleg left. It took ten days and cost 13 lives, but Gene Hambleton was rescued by a US Navy SEAL and a South Vietnamese Commando on the night of April 12, 1972.

The rescue of Capt. Roger Locher is less well known but also demanded extraordinary heroism from a team of rescuers. Locher was a Weapons Systems Officer (WSO) in an F-4. On May 10, 1972, he and pilot Capt. Bob Lodge provided cover for strike flights heading into North Vietnam.

A team with three MiG Kills, Lodge and Locher were shot down in a dogfight.

Lodge, one of the leading American fighter tacticians in Vietnam, was killed, but Locher ejected safely. Unfortunately, he landed near a North Vietnamese air base. Realizing that rescue from this location was nearly impossible, Locher evaded the enemy for three weeks, seeking a better location. Seeing a flight of F-4s fly over him, he raised them on his survival radio.

Rescue forces over northern Laos responded. Two A-1s and two HH-53C Super Jolly Green Giants made an attempt, but enemy fire was too heavy. The next day, the Seventh Air Force commander, Gen. John Vogt, ordered that all strike and support missions on June 2 would support the rescue effort.

Skyraider pilot Capt. Ron Smith led the rescue, but over 100 aircraft were involved. Smith's flight escorted HH-53 pilot Capt. Dale Stovall and his crew, who rescued the downed flier. Smith and Stovall were awarded the Air Force Cross for this, the deepest rescue mission flown into North Vietnam. Gen. Vogt said, "The one thing that keeps our boys motivated is the certain belief that if they go down, we will do absolutely everything we can to get them out." Knowing that the Air Force will do all it can to rescue a lost Airman sustains and strengthens Airmen today.

Operation Arc Light: On June 18, 1965, SAC B–52s flew their first combat missions. The bombers flew so high that they were not seen or heard from the ground, and the bombing power of a three-ship cell might destroy everything

B-52s dropping conventional bombs bring shocking power to battle

in an area one mile long and a half mile wide, shocking the enemy and devastating morale. Many of these strikes were of questionable value, hitting obscure targets in the jungle; as one crewman said, "like swatting flies with a sledgehammer." There were dramatic moments: B-52s wrecked enemy forces besieging Khe Sanh in 1968 in Operation Niagara, with bombs coming down within a half mile of Marine positions. More than 126,000 B-52 sorties were flown over Southeast Asia. During the war, the US lost 31 B-52s: 18 were lost from hostile fire, mostly during Linebacker II, and 13 in mishaps.

As President Nixon applied pressure on North Vietnam in 1972, B-52s stepped up their attacks. SAMs damaged bombers, but none were lost until November 22, 1972, when a B–52 was hit on a mission over North Vietnam. The crew members ejected over Thailand and were rescued. Linebacker II kicked off on December 18, 1972. B-52s struck vital targets in Hanoi and Haiphong in the largest strategic bombing campaign since World War II. During the operation, 741 B-52 sorties were launched; only 12 did not strike their targets. At the same time, more than 200 B-52 strikes were conducted against targets elsewhere. Fifteen B-52s were shot down in this operation. The losses included 33 SAC crewmen killed in action and 33 more taken prisoner. This all-out effort severely impacted North Vietnam's ability to wage war and brought the Communists back to negotiations, and a peace deal on January 23, 1973.

The LAVELLE AFFAIR: John D. Lavelle was a flying instructor and a fighter pilot in World War II. He rose fast when Air Force Secretary Harold Brown selected him to run Igloo White. This project used seismic and acoustic sensors to detect truck traffic on Laos's Ho Chi Minh Trail. This was one of the "looks good on paper" ideas seized upon by Secretary of Defense McNamara. Despite opposition from senior Air Force officers, Lavelle and Brown pressed forward with this, which did not make Lavelle popular among his fellow generals. Secretary Brown helped him rise to four stars as Seventh Air Force commander in South Vietnam, in charge of Air Force operations in Southeast Asia, on July 29, 1971. At the time, rules of engagement banned US aircraft from firing at targets in North Vietnam unless fired upon or engaged by enemy radar. Lavelle recognized that this endangered crews and sought permission to hit SAM sites and MiG bases.

In November 1971, JCS Chairman Admiral Thomas Moorer approved Lavelle's request to attack an enemy air base. Defense Secretary Melvin Laird told Lavelle to take "a liberal interpretation of the rules of engagement." Lavelle ordered his men to hit SAM sites, even those under construction. At the same time, reconnaissance showed that the North Vietnamese were concentrating forces near the Demilitarized Zone (DMZ) that separated North and South Vietnam—preparing a spring offensive. Lavelle viewed this activity—improved and networked air defense radars, troop buildups, increased MiG activity— as meeting all criteria for the protective reaction strikes he was authorized to perform. He ordered a strike to preempt a MiG attack on B-52s bombing the Ho Chi Minh Trail in Laos. The North Vietnamese deployed forces to Dong

Hoi airfield near the DMZ. Lavelle planned to hit the field after they took off, leaving them nowhere to go, while US forces would intercept the aircraft.

Bad weather kept the North Vietnamese grounded, but Lavelle ordered the runway bombed and told the attack commanders—Colonels Charles Gabriel and Jerome O'Malley (both, future four-stars)—to assume that they had an enemy reaction. In the future, flight crews reported "hostile enemy fire" whether they were fired upon or not. They should have stuck to reporting radar or SAM signals. They were permitted to attack enemy air defenses, missile sites, missile support equipment, airfields, AAA guns and radars. However, intelligence officers began falsifying after-action reports to indicate enemy action.

A campaign was launched against the buildup of enemy forces, but no offensive began in the first two months of 1972. Reporters began to question the purpose of the strikes. Simultaneously, the embellished post-mission reports were forwarded by an Airman to his Senator, stating that "we have been falsifying targets struck and bomb damage assessments." Chief of Staff Gen. John D. Ryan sent the Air Force Inspector General to investigate. He concluded that Lavelle had exceeded his authority, violated the Rules of Engagement (ROE) and permitted false reporting of missions as protective reaction strikes when there had been no enemy fire. Ryan offered Lavelle demotion or immediate retirement as a three-star. Defense Secretary Melvin Laird would not speak to Lavelle, so he accepted retirement.

It wasn't over. Hearings in the House of Representatives began on June 12, 1972. Lavelle owned up to his interpretation of the ROE, stating that he meant that Americans went after threats to their forces, not expansion of the war. He pointed out that no civilian areas were hit, and no American planes or lives were lost in these strikes.

Queried about the false intelligence reports, General Ryan said that Lavelle's instructions were behind them. Lavelle assumed responsibility for reporting the strikes as a protective reaction but stated that he was unaware of the four falsified after-action intelligence reports: "I did not do it and did not have any knowledge of the detail. It was my command, and I should have known." Ironically, by the time Lavelle was being grilled, American forces were fighting a no-holds-barred air campaign over North Vietnam.

John Lavelle took the brunt of this awful situation, in which political figures made rules that placed aircrews at risk and punished Airmen who tried to protect their men. Lavelle retired, his reputation ruined by news outlets and treated like a wild-eyed maniac bent on fighting his own war. Congress ordered him retired as a major general, one more demotion.

Sadly, this wasn't the only time leaders were caught in a lie. Cambodia was off limits but was heavily bombed, with knowledge of the facts limited to a handful of commanders. The bombardment of Cambodia began in 1969, under the control of PACAF (Pacific Air Forces). Its commander: Gen. John D. Ryan.

In 2007, after reviewing tapes from the Nixon White House, retired Lt. Gen. Aloysius Casey concluded that this previously unavailable evidence "fully vindicates the truthfulness of General John D. Lavelle before the United States Congress." Author Seymour Hersh, a leading Lavelle critic in 1972, agreed that Lavelle had done what the president wanted.

In a 1978 interview, Lavelle said, "We just should have said, 'Hey, either fight it or quit, but let's not waste all the money and the lives the way we are doing it.'" Lavelle died in 1979. His son described him as brokenhearted. In so many ways, the Vietnam War is a disturbing chapter in American history.

POWS

The extraordinary courage of the POWs in Vietnam was one of the few rallying points for the general public. Little publicity had been given to the heroes of this war, and almost in recompense, the POWs received great adulation.

George "Bud" Day: exemplar of heroism

George E. "Bud" Day is the most decorated Airman in history. A Marine in World War II and a fighter pilot in Korea, he was captured on August 26, 1967, after the North Vietnamese shot down his F-100 just north of the DMZ. Severely wounded, he was taken to a POW camp, interrogated and tortured. Despite terrible injuries, Day escaped and evaded for two weeks. He crossed the DMZ and got within two miles of a US Marine outpost when he was shot and recaptured by the Viet Cong. In poor health, he endured beatings, torture and starvation while continuing to resist his captors until he was released in 1973. He retired from the Air Force in 1977, died in 2013, and was posthumously promoted to brigadier general in 2018.

Lance P. Sijan was a 1965 Air Force Academy graduate. He arrived at Da Nang Air Base, Vietnam, in July 1967. On November 9, on a strike mission against targets on the Ho Chi Minh Trail in Laos, Sijan and his squadron commander were making their attack when their F-4C exploded, possibly hit by ground fire. His pilot, Lt. Col. John Armstrong, was never found. Sijan managed to eject and, despite severe injuries—a compound fracture of his left leg, severe hand and head injuries and lacerations—he crawled for six weeks, evading the enemy, seeking to return to friendly forces. He eventually passed out and was found by North Vietnamese soldiers. Even in his emaciated and battered state, he escaped. Quickly recaptured, he was tortured and transported to Hanoi, a journey of three nights in an open truck in cold rain. Weakened by his ordeal, Sijan contracted pneumonia and died. He was 25 years old. Captain Sijan received the Medal of Honor posthumously.

One attempt was made to rescue the POWs. The Son Tay Raid on November 21, 1970, saw a Special Forces team attempt to rescue Americans from Son Tay POW camp 20 miles west of Hanoi. Unfortunately, the prisoners had been moved elsewhere. A heroic effort, but it went for nought.

Operation Homecoming was the return of 591 US prisoners of war. On February 12, 1973, the first of several flights brought released American prisoners of war from Hanoi, North Vietnam, to Clark Air Base in the Philippines. The last Vietnam POW serving in the Air Force, Maj. Gen. Ed Mechenbier, flew back to Hanoi in 2004 to retrieve the remains of Americans, recovered by the Vietnamese. The C-141 used on that flight was the one that carried the first POWs out of Hanoi in 1973.[14]

An enduring part of the POW legacy is the tap code used to communicate between cells. Rather than speaking, rapping one's hand against a wall in a pattern could pass information without drawing the enemy's attention. Much "shorthand" was used to facilitate this. One of these messages is still used widely among Airmen: GBU. It's short for "God Bless You," and is often used when someone is deployed or facing personal illness or difficulty.

The last mission of the war in Southeast Asia was flown on July 15, 1973, by an A–7D Corsair II. Since 1962, the Air Force had flown more than five million sorties over South Vietnam, North Vietnam, Laos and Cambodia, losing 2,251 aircraft: 1,737 to hostile action and 514 to other operational causes. About 2,586 Airmen lost their lives in the war.

Operation Babylift took place in April 1975. Recognizing that orphans and children fathered by Americans would face a miserable life after a Communist takeover, the US tried to evacuate as many as possible. The operation began tragically as a C-5 full of children crashed near Saigon. Miraculously, 175 of the 313 people on board survived the accident. After this, more than 3,300 children were safely removed from Vietnam. There were many heroes in that terrible crash on April 4, 1975: the pilots, Capts. Dennis Traynor and Tilford Harp, were both awarded the Air Force Cross for recovering the airplane after an explosive decompression blew out a cargo door and wrecked the hydraulics and flight controls. After the crash, Flight Nurse First Lt. Regina Aune demonstrated incredible courage. Thrown the length of the cargo compartment by the impact, she got to her feet and began carrying children out of the wreckage to safety. She did this despite a broken vertebra, leg

---

[14] This airplane is in the Air Force Museum.

and foot, and severe lacerations. She was awarded the Airman's Medal and became the first woman to receive the Cheney Award, given for bravery in a humanitarian air mission.

EVACUATION

At about the same time, the US began the evacuation of Phnom Penh, Cambodia. Almost 900 Cambodians were flown to Thailand aboard USAF C–130s. As the situation in South Vietnam became dire, the Air Force stepped up evacuation efforts, moving 45,000 people out of Saigon. By April 29, fixed-wing aircraft could no longer use Tan Son Nhut Airport, so helicopters (Marine, Navy, and Air Force) took over the mission. Six thousand more people were evacuated in two days. Saigon fell on April 30, 1975. Military Airlift Command (MAC) airlifters were engaged in humanitarian airlift for much of 1975, moving refugees from Southeast Asia to Guam and the mainland United States.

On May 12, 1975, Cambodian gunboats seized the US merchant ship Mayaguez and its 40-man crew, off the Cambodian coast. In response, MAC brought Marines and equipment from the Philippines and Okinawa to Thailand. The crew of the Mayaguez was rescued unharmed, but the Marines found themselves in a pitched battle with Cambodian Khmer Rouge troops on Koh Tang Island, supported by USAF helicopters and AC-130 and A-7 aircraft. Two Airmen and 15 Marines were killed. It summarized the war: Americans had fought bravely in a struggle that brought their country nothing.

CHAPTER 7
COLD WAR AND A DANGEROUS WORLD
1970s TO 1991

"Of the four wars in my lifetime, none came about because the United States was too strong."
- Ronald Reagan

During my youth, in the wake of Vietnam, I feared that honor and patriotism were of the past. Our country changed because of the trauma of Vietnam, and the deceit that cost so many lives.

Ronald Reagan told America that we must stand for freedom. We could use our economic power, our freedom and our moral strength to strike at the rule of the Soviet Communists, under the shield of our military might. Within a decade, the USSR collapsed—outspent, outmaneuvered and outproduced by liberty and capitalism. Soviet satellites in Eastern Europe chose freedom. The words of Reagan's first inaugural speech stay with me:

"Freedom and the dignity of the individual have been more available and assured here than in any other place on earth... The price for this freedom at times has been high, but we have never been unwilling to pay that price. As for the enemies of freedom, those who are potential adversaries, they will be reminded that peace is the highest aspiration of the American people. We will negotiate for it, sacrifice for it; we will not surrender for it—now or ever, together, with God's help, we can and will resolve the problems which now confront us. And after all, why shouldn't we believe that? We are Americans."

Those words shaped my life.

PEACE THROUGH STRENGTH

Two of the Air Force's best aircraft came into service after Vietnam. McDonnell-Douglas was named prime contractor for the USAF's new air-superiority fighter in December 1969. The first F-15A was delivered to the 1st Tactical Fighter Wing in January 1976. The Eagle set numerous time-to-climb records and was the first fighter to have thrust greater than its weight, allowing it to accelerate while going straight up. The F-15E, designed for ground attack, first flew in 1986. The F-15 established a 100-0 air-to-air victory ratio in US and foreign service. The F-15E is still a primary ground attack platform.

In January 1975, Air Force Secretary John McLucas authorized purchase of the General Dynamics F-16, a low-cost, lightweight, highly maneuverable aircraft to supplement the more costly F-15. The first Fighting Falcon (or Viper, as Airmen call it) came into the fighter force in early 1979. The F–16 proved versatile, able to perform Wild Weasel, strike and air-superiority missions. Many allied nations chose to buy the F-16 in the 1980s and 1990s. The Viper is still in wide use and has seen action in all the conflicts fought by US forces since 1990, with an outstanding combat record.

As the Air Force renewed its fighter fleet, it also introduced a new tanker. The first McDonnell-Douglas KC–10A Extender arrived in 1981, and more were being delivered. Substantially larger than the KC–135, the Extender could not only carry nearly twice the fuel and cargo, but also easily refuel more types of aircraft, with an installed probe and drogue. At the KC-135 crew training unit in 1984, an instructor told me to seek transfer into another aircraft: "The -135 is thirty years old, it will be replaced soon." At the time, this seemed reasonable but the KC-10 was costly, and only 60 were purchased. The last KC-10 was retired in September 2024. I, too, retired, but the KC-135 continues to fly.

Despite this new technology, America after Vietnam was unsure of itself. Not surprisingly, challenges arose. In November 1979, Iranians invaded the US embassy in Tehran and held 66 Americans hostage because the US refused to hand over the deposed Shah of Iran. Operation Eagle Claw was an attempt to rescue these Americans in April 1980. Mechanical problems rendered two helicopters non-mission capable, then a Marine RH-53 collided with an Air Force EC-130 at a refueling site in the Iranian desert. This caused an explosion that killed eight men and destroyed the two aircraft. President Jimmy Carter aborted the mission at that point. The team departed on the remaining MC-130s, abandoning the remaining four helicopters and the eight dead Americans in the Iranian desert. On January 20, 1981, two C–9 Nightingales transported the Americans released by Iran from Tehran to Rhein-Main AB, Germany. A few days later, they returned to the United States.

America tried to be a peacekeeper. In 1982, President Ronald Reagan sent US Marines into Lebanon, with British, French and Italian troops. After a terrorist bombing by Iranian proxy Hezbollah killed 241 Marines, sailors and soldiers on October 23, 1983, the Air Force evacuated the dead and 95 wounded Americans. The peacekeeping forces pulled out in early 1984.

The Cold War continued. On September 1, 1983, a Soviet Su–15 shot down a Korean Air Lines Boeing 747 with 269 people aboard near Sakhalin Island, north of Japan. The Soviets claimed it was a spy plane, claimed that they mistook it for an RC-135 reconnaissance aircraft, and generally lied and postured to such an extent that the world was disgusted by their perfidy and brutality.

The Air-Launched Cruise Missile (ALCM) demonstrated the value of a standoff weapon. In January 1981, Boeing delivered the first ALCMs, which could deliver a nuclear weapon to a target 1,500 miles away, enhancing the reach of the B-52. By 1990, the Conventional ALCM (CALCM) was available, and was used in conflicts since then. ALCMs have been replaced by the Advanced Cruise Missile (ACM), and in the conventional role by the Joint Standoff Weapon (JSOW).

One of the USSR's gambits was clandestine involvement in revolutionary activities. In October 1983, following a Communist coup, the US and several Caribbean states invaded Grenada to evacuate Americans and restore political stability. SAC tankers, TAC fighters, and MAC airlifters supported the quickly won operation with paratroop drops and close air support.

Libya was a Soviet ally and a supporter of global terrorism. Following several terrorist attacks, culminating in a bombing at a disco in Germany which killed three people, including two US military members, Operation Eldorado Canyon was ordered. This was a retaliatory attack on Libyan dictator Muammar Qadhafi and his military. On April 14, 1986, 24 F-111 fighter-bombers, supported by 28 KC-135s and KC-10s, launched from bases in England. In attacks coordinated with naval forces in the Mediterranean, several targets were hit. Qadhafi survived and remained in power, albeit less confrontational. One F-111 was shot down, its two-man crew killed in the operation.

The Middle East remained a flashpoint. Air Force missions included E-3 AWACS aircraft to monitor air traffic and KC-135 tankers to support Navy fighter aircraft in this volatile environment. The first E-3A Sentry aircraft arrived at Tinker AFB, Oklahoma, in 1977. The Sentry carried a large rotating radar disk above its fuselage. It was the successor to the Vietnam-era EC-121, known as Red Crown in Southeast Asia. E-3s served in Saudi Arabia in 1979, then in Operation Elf beginning in October 1980. This deployment of four AWACS and three KC-135s to Riyadh was to protect Saudi airspace during the Iran-Iraq War. With political unrest in Poland in 1981, AWACS planes deployed to Germany, and then to Egypt following the assassination of Prime Minister Anwar Sadat. NATO created its airborne early warning force at Geilenkirchen NATO AB, Germany, in February 1982. Operation Earnest Will protected oil tankers in the Persian Gulf from July 1987 through September 1988. When Iran placed mines in the Gulf, C-5 and C-141 aircraft airlifted minesweeping personnel and equipment to the area.

Through all this turmoil, the US and USSR kept negotiating. On December

8, 1987, the two signed the Intermediate-Range Nuclear Forces (INF) Treaty. This removed US Pershing II and Ground Launched Cruise Missiles (GLCM), and Soviet SS-20 intermediate-range missiles from Europe. There were six Air Force GLCM wings in Europe: two each in Britain and the Netherlands, and one each in Germany and Italy. The 501st Tactical Missile Wing at RAF Greenham Common in Britain was the first GLCM unit activated, on July 1, 1982, and the last closed on May 31, 1991.

A special, though strained, relationship existed between the United States and Panama due to the American role in Panamanian independence and the Panama Canal Zone, which bisected the isthmus. Most Panamanian leaders sought good relations with the Americans. But strongman Manuel Noriega was involved with drug trafficking. When the US supported regime change through electoral processes, Noriega reacted with harassment, assault and murder. The American garrison in Panama was augmented in April 1988 by 1,200 additional soldiers. When Panamanian forces murdered a Marine lieutenant in December 1989, President George H. W. Bush decided to remove Noriega. The US invasion of Panama began on December 20, 1989, just after midnight local time. The operation involved 27,000 US troops and over 300 aircraft, including AC-130 gunships, OA-37B attack aircraft, C-130, C-141 and C-5 airlifters, KC-135s, F-117A Stealth aircraft and AH-64 Apache helicopters.

US attacks struck the airport, military posts, barracks, command centers and Noriega's residence. Navy SEALs destroyed Noriega's private jet and sank a Panamanian gunboat. The F-117A Stealth Fighter saw its first action in Panama.[15] Two F-117s dropped 2,000-pound laser-guided bombs on the Panama Defense Forces (PDF) barracks at Rio Hato, then C-130s dropped Army Rangers to seize the enemy base. An AC-130H destroyed the heavily defended PDF headquarters in Panama City. Nearly 10,000 airborne troops from Pope AFB parachuted into Panama from C-141s, the largest night-combat airdrop since D-Day in 1944. Primary combat operations ended within a few days, though US troops engaged hostile forces into January 1990.

---

[15] The futuristic-looking little jet would be a mainstay of operations in Iraq, Bosnia, Kosovo and Afghanistan.

CHAPTER 8
BOOKS ABOUT THE AIR FORCE

"My wings are a thousand books."
- Gill Robb Wilson, World War I pilot,
writer and a founder of the Civil Air Patrol

There are thousands of books about the Air Force in all its aspects: heroes, aircraft, campaigns, strategies, tactics, bases, missiles, missions—you name it. What follows are some recommendations.

Beyond the Wild Blue by Walter Boyne. Boyne was a SAC bomber pilot throughout his Air Force career, and he wrote hundreds of articles and 50 books. This is a survey of the Air Force from 1947 to 2007.

Hostile Skies by James J. Hudson. This combat history of the American Air Service in World War I is well-written and thorough. It's an excellent book.

Fighting the Flying Circus is Eddie Rickenbacker's memoir of World War I aerial combat. Capt. Rickenbacker was brutally honest and straightforward, qualities that come through in this book.

Billy Mitchell by Alfred Hurley; *A Few Great Captains* and *Forged In Fire* by DeWitt S. Copp. These are vital to understanding the development of the Army Air Forces into the global colossus that defeated enemy nations.

Hap Arnold and the Evolution of Airpower by Dik Daso. Hap Arnold's vision of what airpower could be and the foundations he laid down for the future of the Air Force are lasting.

I Could Never Be So Lucky Again by Jimmy Doolittle and C.V. Glines. Follow this with *The Doolittle Raid* by Glines. An Air Force pilot and writer, Glines covered the Doolittle Raiders so thoroughly that they made him an honorary Raider. Doolittle's autobiography is exciting. The man led an extraordinary life. The Doolittle Raid story has never been told better than by Glines.

Mission With LeMay by Curtis E. LeMay and MacKinlay Kantor; pair this with Warren Kozak's *LeMay*. Few military figures are as controversial as General LeMay. Blunt and deadly serious, he shaped tactics in the European war and the conflict's outcome in the Pacific. In the Cold War, he dedicated the modern Air Force to power for peace, maintaining the strength to deter and, if necessary, destroy an enemy. His influence is still felt in the Air Force.

Black Snow by James M. Scott, and *The Bomber Mafia* by Malcolm Gladwell. These books describe the bombing of Japan. Gladwell delves into the personalities of the leaders and their doctrines, while Scott digs deep into strategies and the experiences of those involved in these horrifying but war-winning actions, both Americans and Japanese.

Masters of the Air by Donald L. Miller. Read this with Bert Stiles' *Serenade to the Big Bird,* or Harry Crosby's *A Wing and A Prayer.* Miller tells the high-level story of why strategic bombardment was chosen to wage war on Germany, how the strategy was planned and how it worked in practice, along with vignettes of Airmen in combat. Stiles' book tells of his adventures as a B-17 pilot in 1944. It's as close as we can come to chatting with a young man in the 1940s. He visits a girlfriend in London and lives a pleasant life when he's not in battle. His descriptions of flight are lyrical, and the sudden terror of air combat is compelling. Stiles completed his tour, transferred into P-51 fighters, and was killed in action in November 1944. Harry Crosby wrote one of the most thoughtful works on flying in the Mighty Eighth. A navigator, he came through some of the worst of the war during 1943-44, when aircrews had little hope of surviving a combat tour. He tells the story with heart and insight.

Forgotten Fifteenth by Barrett Tillman. "It's still the same old story; the Eighth gets all the glory, as flak goes by." That bit of doggerel (set to "As Time Goes By" from the film *Casablanca*) illustrates why this book is on this shelf. Tons of books cover the Mighty Eighth, but the Fifteenth Air Force, which supported the brutal Italian campaign and struck targets in southern France, Germany, Austria, Czechoslovakia and the Balkans, receives little recognition. The Fifteenth did an impressive job under challenging circumstances.

Flying With the Flak Pak by Kenny Kemp is a scrapbook of the author's father's days in the Army Air Forces as a B-24 pilot in the Southwest Pacific. Kemp shows day-to-day life in a seldom-remembered theater of war, as men fly an airplane that rarely gets its due. It is a fun read, shining light on a part of the war that has not been told much.

Winged Victory by Geoffrey Perret. There are more "academic" works on the Army Air Forces in World War II, but this is a readable and thorough single-volume history for the general reader. It is an excellent summary.

Officers in Flight Suits by John Darrell Sherwood, and *Hot Shots* by Jennie Ethell Chancey and William R. Forstchen. These are social history and tales of derring-do in the skies over Korea. The "flight suit" culture is personified by fighter pilots, but is part of every Airman wearing silver wings.

By Any Means Necessary by William Burrows, and *Silent Warriors, Incredible*

Courage by Wolfgang Samuel depict Cold War reconnaissance missions. Burrows focuses on operations and taskings, while Samuel describes what it was like to fly these harrowing missions at a time when the United States government might not even acknowledge your presence or purpose, even if you were attacked by a hostile nation under surveillance.

Thud Ridge by Col. Jack Broughton, and *When Thunder Rolled* by Maj. Ed Rasimus. These two books will enthrall and enrage the reader. American Airmen fought with awe-inspiring bravery over North Vietnam, to no avail. Sometimes, even shooting back at their enemies was forbidden. Broughton was a vice wing commander whose frustration with the command structure comes through loud and clear. Rasimus was a lieutenant, fresh from flight training, thrown into some of the most ferocious air combat of the war. He, too, describes frustration and anger stemming from unreasonable rules of engagement that gave American enemies every advantage while placing American Airmen at incredible risk.

The Eleven Days of Christmas by Marshall L. Michel. The "Linebacker II" campaign against North Vietnam in December 1972 still sparks controversy. Michel, an F-4 pilot who flew in Vietnam, has produced an outstanding book that cuts through the controversies about tactics and some of the legends (mutiny, mass refusals to fly) surrounding this operation.

The Passing of the Night by Robinson Risner; *Into the Mouth of the Cat* by Malcolm McConnell; *American Patriot* by Robert Coram. The stories of Robbie Risner, Lance Sijan and Bud Day are told throughout the Air Force as examples of the heroism and horrors of war, both in actual combat and then in the hands of the enemy as prisoners in North Vietnam. Shocking and inspiring, they demonstrate the costs of wearing wings in desperate battles, both in the advanced world of a fighter cockpit and in the filth and degradation of a prison cell in a hostile land.

Fighter Pilot by Christina Olds, Robin Olds and Ed Rasimus. The embodiment of the term "fighter pilot," Robin Olds led a life that would be unbelievable if it were fiction: All-American football player, flying ace, and a wing commander whose reputation took him to the Pentagon and the White House, where his lessons on the war in Vietnam went unheeded. A maverick and a warrior, Robin Olds defined leadership.

The Right Stuff by Tom Wolfe, and *Riding Rockets* by Mike Mullane. *The Right Stuff* catapulted Chuck Yeager and the Mercury astronauts back into the headlines in the early 1980s and remains riveting. *Riding Rockets* is a look inside the astronaut experience, a peek into a flying squadron, albeit an elite one. It's eye-opening, funny, occasionally crude and always honest about the people and the bureaucracy of the space program. It is the best astronaut memoir.

The Mind of War by Grant Hammond, or *Boyd: The Fighter Pilot Who*

Changed The Art of War by Robert Coram. Col. John Boyd was an iconoclast, a rebel who rose to the rank of colonel despite his personal quirks and a total lack of tolerance for anything—bureaucracy, leaders, or politicians—that did not meet his standards. Hammond delves into the philosophies promoted by Boyd, while Coram focuses more on the man. Much of Boyd's thinking is innovative and offers new paradigms for military strategy. Some of it is just one man's fiercely held opinion.

More Than a Tanker by Robert S. Hopkins III. There are countless books about specific aircraft, equipment, uniforms, etc. Much is written about almost all the aircraft in the Air Force inventory, and readers will choose their favorites. I chose this book because it covers a subject I know well—the airplane I flew. Books of this type must thoroughly cover history, technology, people and mission. Hopkins does this expertly. This is the best history of the -135 family.

The Limits of Airpower by Mark Clodfelter. While bombardment is essential to the Air Force's mission, it is not a cure-all that can win when an overall national strategy is flawed. Clodfelter explains why airpower may not achieve military or political objectives in a limited war.

CHAPTER 9
CELEBRITIES IN AIR FORCE BLUE

"If I did half the things [Williams] did in that movie, I'd still be in Leavenworth."[16]
- Adrian Cronauer

MEDIA

Sunny Anderson—The Food Network star was an Air Force broadcaster on AFN, stationed in South Korea and Texas during her service.

Art Bell—A radio host specializing in paranormal phenomena, Bell was an Airman during the early years of the Vietnam War.

Adrian Cronauer—The subject of *Good Morning Vietnam*, he was an Airman who went on to be an attorney and a veterans' advocate.

James Dickey—The author of *Deliverance* and a Poet Laureate of the United States served as a radar operator in P-61 night fighters in World War II.

Ernest K. Gann—One of the greatest aviation authors, Gann started on Broadway and flew as a hobby. In 1938 he took a job with American Airlines flying DC-2s. In World War II, he was in the AAF's Air Transport Command, flying what today would be called strategic airlift, hauling passengers and cargo across the Atlantic, Africa and India.

Joseph Heller—The author of *Catch-22* flew 60 missions as a B-25 bombardier in Italy.

Don Herbert—TV's *Mr. Wizard* taught science to children, but before that he was a B-24 pilot in the Fifteenth Air Force, earning a DFC.

Stacy Pearsall—This photojournalist was an Air Force Combat Camera troop decorated for bravery in Iraq.

Robin Quivers—Howard Stern's radio co-host was an Air Force nurse for about three years in the 1970s.

---

[16] The U.S. military prison is at Fort Leavenworth, Kansas.

Bob Ross—The gentle-voiced teacher who brought the joy of painting to millions spent 20 years in the Air Force, including a tour as a Military Training Instructor and finished up as a First Sergeant at Eielson AFB, Alaska.

Bob Schieffer—The CBS reporter was an Air Force officer from 1959 to 1962.

Hunter S. Thompson—The Gonzo journalist wrote for an Air Force base newspaper. His talent was apparent from the beginning, but he got an early-out honorable discharge, because "this Airman, although talented, will not be guided by policy."

Thornton Wilder—The author of *Our Town* and *The Bridge of San Luis Rey* served as an intelligence officer in North Africa and Europe in World War II.

Howard Zinn—This historian, whose *A People's History of the United States* influenced generations of students, was a B-17 bombardier with the 490th Bomb Group of the Eighth Air Force.

MUSIC

Gene Autry—The singing cowboy who popularized "Rudolph the Red-Nosed Reindeer" flew C-109 transports (modified B-24s), hauling gasoline over the Hump to forces in China.

Samuel Barber—The composer spent three years in the AAF in Special Services, providing entertainment for troops.

Boxcar Willie—The country singer was a flight engineer on several types of aircraft in a 27-year USAF/ANG career, retiring as a master sergeant in 1976.

Johnny Cash—The "man in black" was a young man in blue who served in an intelligence unit in Germany in the 1950s. It's been reported that he was the first person outside of the USSR to hear of Josef Stalin's death.

Tennessee Ernie Ford—Famous for "Sixteen Tons," the country singer was a bombardier on B-29s in combat over Japan in 1945.

Don Ho—Before he became Hawaii's musical ambassador, the singer was an Air Force pilot flying C-97 cargo planes in the late 1950s.

Glenn Miller—The Big Band leader volunteered for the AAF in 1942. James Stewart played him in a 1954 biopic.

Willie Nelson—He joined in 1951 and was out on medical discharge in 1952. I have no word on any

Glenn Miller

use of medicinal drugs.

Mike Nesmith—The Monkees' ski-capped guitarist was an aircraft mechanic in SAC in the early 1960s.

Mel Tillis—The country music star was a baker in the Air Force, 1951 to 1955.

John Williams—The film composer was a USAF Bandsman from 1951-55, a pianist, and arranger. His film scores include *Star Wars, Saving Private Ryan, Schindler's List* and approximately 100 more movies, as well as TV music from *Wagon Train* to *Meet the Press*.

PERFORMING ARTS

Robert Altman—A much-honored film director, he flew 50 missions in B-24s in the South Pacific.

Martin Balsam—An award-winning actor who played hundreds of roles, from *12 Angry Men* to *The Twilight Zone*, was stationed in the China-Burma-India Theater during World War II.

Edward Binns—A character actor in *12 Angry Men, Fail-Safe, Judgment at Nuremberg, Patton* and more than 500 TV episodes, he also served in the CBI.

Charles Bronson—The toughest guy in the movies was a B-29 gunner in World War II. He flew 25 missions against Japan and was wounded in action.

George Carlin—Unsurprisingly, this comedian didn't fit into the Air Force mold in the 1950s. He received a general discharge following several "incidents."

Jackie Coogan—A child star in silent movies and beloved as Uncle Fester on *The Addams Family*, he was a glider pilot in the CBI.

Sam Elliott—The actor was an ANG member in the 146th Airlift Wing. He would've won the Mustache March competition, hands down.

John Frankenheimer—The director of *The Manchurian Candidate* and *Seven Days in May* was assigned to an Air Force film unit in the 1950s.

Morgan Freeman—The smooth-voiced actor enlisted in the Air Force and was a radar repairman during his four years in service, 1955 to 1959.

Clark Gable—The man who played Rhett Butler flew in combat aboard B-17s over Nazi-occupied Europe while making training films in 1943.

Larry Hagman—When he became Major Tony Nelson in *I Dream of Jeannie*, Hagman already had worn Air Force blue as an enlisted man from 1952 to 1956.

Howard Hawks—The Oscar-winning director was a pilot during World War I, which likely influenced his several aviation-themed movies.

Sherman Hemsley—The star of TV comedies *The Jeffersons* and *Amen* spent four years in the Air Force from 1955 to 1959.

Charlton Heston—This Academy Award recipient served as a radio operator and gunner on B-25 bombers in the Aleutians during 1944-45.

John Hillerman—A recipient of an Emmy Award for his role in *Magnum PI*, Hillerman maintained B-36s during his 1953-57 Air Force stint.

Russell Johnson—The professor on *Gilligan's Island* had real-life experience of being lost on a tropical island. He was shot down on his 44th mission as a B-25 bombardier in March 1945, off the coast of Mindanao in the Philippines, and was rescued. No word on whether he encountered Ginger or Mary Ann.

Norman Lear—The producer of many television comedies, including *All in the Family, Maude, Sanford and Son, The Jeffersons* and *Good Times,* flew 52 missions as a radio operator in B-17s with the Fifteenth Air Force.

Gordon MacRae—The star of movie musicals *Oklahoma* and *Carousel* was a navigator in C-47s during World War II.

Delbert Mann—An Academy Award-winning director, he flew 35 B-24 missions in the Eighth Air Force in World War II. He directed *A Gathering of Eagles*.

Walter Matthau—An Academy Award recipient who starred in films from *The Odd Couple* to *Grumpy Old Men*, he was a gunner in James Stewart's 453rd Bombardment Group, flying B-24s against Germany.

Jerry Mathers –The *Leave it to Beaver* star served in the Air National Guard from 1966 to 1969.

Chuck Norris—*Walker, Texas Ranger* star Chuck Norris first wore a badge as an Air Force Air Police troop. Norris learned his martial arts skills while stationed in Korea. When people tease about the Air Force not being so tough, remember: Chuck Norris is an American Airman.

Tom Poston—For more than fifty years, if you watched TV sitcoms you were likely to see Tom Poston. He flew C-47s and dropped paratroopers on D-Day.

Gene Roddenberry—The *Star Trek* creator flew 89 missions in the Pacific as a B-17 pilot and earned a DFC.

Dan Rowan—A comedian best remembered as co-host of *Laugh-In*, Rowan was a fighter pilot in World War II. He shot down two Japanese aircraft before being severely wounded himself. He was awarded two DFCs.

Roy Scheider—The star of *Jaws* and *All That Jazz* was an Air Force officer from 1955 to 1958.

Sinbad—The comedian was a boom operator at McConnell AFB for a few years before deciding that the Air Force wasn't for him. Unfortunately, he still had time on his enlistment and left the Air Force on less than friendly terms.

James Stewart—An Academy Award winner who starred in movies from the 1930s to the 1970s, Stewart was a private pilot who volunteered for the AAF in February 1941. He flew combat missions in B-24s with the Eighth Air Force, receiving two DFCs, and stayed active in the Air Force Reserve until 1968, retiring as a brigadier general. Look at Stewart's boyish face in 1942's *Winning Your Wings* and contrast it with the haggard George Bailey he portrayed in 1946's *It's A Wonderful Life* to see the changes war can bring in a man. It is said that the scene where George tearfully prays for guidance reveals Stewart's stress from the war. James Stewart was an honored professional in his craft, a patriot and by all accounts a good guy. An American hero.

James Stewart: leader by example

Fred Ward—He is remembered for his portrayal of Gus Grissom in *The Right Stuff*, and he was a radar technician during his enlistment in the early 1960s.

William Wyler—One of the greatest directors in Hollywood history, Wyler was nominated for 12 Oscars, winning three times, for *Mrs.Miniver, The Best Years of Our Lives,* and *Ben-Hur*. In World War II, he made one of the greatest aviation films, *Memphis Belle: A Story of a Flying Fortress*. He flew combat missions to obtain the footage needed for the film, earning an Air Medal but suffering significant hearing loss.

POLITICS

Ben Nighthorse Campbell—This senator and 1964 Olympian in judo (three-time US champion) is a Korean War veteran who served in the Air Police.

Bob Dornan—This controversial California congressman flew fighters on active duty and then in the California ANG for over twenty years. In Congress, he was a vocal advocate for airpower.

Robert Gates—Secretary of Defense under Presidents Bush and Obama, he was an Air Force intelligence officer from 1967-69, then moved to the CIA.

Barry Goldwater—A US senator from Arizona, he was the 1964 Republican candidate for president. He was a cargo pilot who flew the Hump. After the war, he was a founding member of the Arizona ANG and rose to major general in the Air Force Reserve.

Fiorello LaGuardia—The congressman and mayor of New York was a World War I pilot flying Caproni bombers on the Italian Front.

George McGovern—Senator and 1972 presidential candidate, the South Dakotan flew 35 missions in B-24s with the Fifteenth Air Force, earning a DFC.

Robert McNamara—Some of his first statistical work was in devising schedules to maximize the efficient use of B-29s. His ideas as Secretary of Defense during the Vietnam War were far less successful.

Ron Paul—A leading figure in American libertarianism, Paul was an Air Force flight surgeon in the 1960s.

Rick Perry—The former Texas Governor was an AFROTC graduate of Texas A&M and a C-130 pilot from 1972 to 1977.

William Rehnquist—The Supreme Court Justice was a weather observer in World War II, serving in the United States and North Africa.

Elliott Roosevelt—Follow this story if you can: He was the son of President Franklin Roosevelt and entered the AAF (despite being classified 4-F, physically unfit for military service) in 1940 as a captain (with no previous military training) and became a pilot (without attending military flight training), and flew 89 combat reconnaissance missions. His official Air Force biography shows him wearing pilot wings but states that he was rated an Aircraft Observer and attended navigator training in 1941. He was highly decorated, a group and wing commander, and was promoted to brigadier general at 34. He also found time to be embroiled in a scandal involving Howard Hughes, Hollywood starlets, bribery allegations and the selection of a Hughes aircraft for the War Department, "against [the] better judgment" of Gen. Hap Arnold and the Army Air Force procurement establishment. Unusual, *n'est-ce pas*?

Quentin Roosevelt—President Theodore Roosevelt's youngest son volunteered for the Air Service at 19 as the US entered World War I. He went into combat as a pilot. He recorded one victory in aerial combat, but just a few days later, on July 14, 1918, he took two German bullets in the head and died behind enemy lines. The Germans buried him with full military honors.

Ted Stevens—He served as a Senator from Alaska for 40 years before losing his seat after a tainted corruption investigation led to a guilty verdict that was thrown out later for prosecutorial misconduct. Stevens held two DFCs from his service as a C-46 pilot in the CBI.

Stewart Udall—a congressman and conservationist, he was a B-24 gunner in the 454th Bomb Group, flew 56 missions in the Fifteenth Air Force.

George Wallace—The segregationist Governor of Alabama was a flight engineer on B-29s in the Pacific.

Jim Wright—A former Speaker of the House, he was a bombardier in B-24s with the 380th Bomb Group in the Pacific, receiving the DFC for his service.

Coleman Young—The longtime mayor of Detroit and Tuskegee Airman qualified as a B-25 navigator and was involved in the Freeman Field Mutiny.[17]

SPORTS

Alonzo Babers—A 1983 USAFA graduate, he won two Olympic gold medals in the 1984 Summer Olympics (400 meters, 4x400 relay), then went to pilot training and served as a pilot in the Air Force, the Reserve and for United Airlines.

Chuck Bednarik—The NFL Hall of Famer who played offense and defense—60 minutes in most games—was a waist gunner in the Mighty Eighth, flying 30 combat missions over Germany.

Vic Carapazza—A Major League Baseball umpire, he served in Security Forces from 1998 to 2002.

Doc Counsilman—He set world records in breaststroke and was a six-time national champion swimming coach at Indiana University. He trained 60 Olympians, including Mark Spitz, and twice coached US Olympic swim teams. In World War II, he was a B-24 pilot in Fifteenth Air Force and earned a DFC.

Elmer Gedeon—Gedeon played for the Washington Senators, then joined the military in 1941 and earned a Soldier's Medal for rescuing crewmen from a crashed airplane. He flew the B-26 in Europe and was killed in action on a mission to bomb German V-1 rocket sites on April 20, 1944.

Ben Hogan—The golfer was assigned in Texas as a utility pilot and instructor.

Bernard James—Before the Dallas Mavericks forward played in the National Basketball Association, he was a Defender in the Air Force and deployed to Kuwait, Qatar and Iraq.

Bobby Jones—The champion golfer volunteered for duty with the Air Corps and went to Europe in 1944-45 as an intelligence officer with Ninth Air Force.

---

[17] In 1945, a group of black Airmen at Freeman Airfield in Indiana were arrested for protesting segregation of officers' clubs on the base.

Micki King—A diver at the University of Michigan, she joined the Air Force in 1966 and finished fourth at the 1968 Olympics despite a painfully broken arm. Assigned as a physical education instructor and diving coach at the Air Force Academy, she returned to the Olympic Games in 1972, winning a gold medal in the three-meter springboard event. Col. Maxine "Micki" King retired from the Air Force in 1992. Her daughter graduated from the Academy in 2004.

Tom Landry—The coach who led the Dallas Cowboys to two Super Bowl titles was a veteran of 30 missions over Germany, piloting a B-17 of the 493rd Bomb Group in the Eighth Air Force.

Buddy Lewis—A baseball star with the Washington Senators—2-time all-star, .297 lifetime average—Lewis served from 1941-45, winning three DFCs while flying C-47s on 368 missions over the Hump.

Shane Livensparger—This Major League Baseball umpire serves as an officer in the Georgia ANG.

Nick Mahrley—A Major League Baseball umpire, he worked in Survival Equipment during his Air Force service.

Gregg Popovich—An Air Force Academy basketball star, he coached the San Antonio Spurs to five NBA titles. He was an assistant coach at the Academy and an intelligence officer.

George Steinbrenner—The New York Yankees owner for thirty years won seven world championships and served as a lieutenant at Lockbourne AFB, Ohio

OTHERS

Carroll Shelby—The car designer was a flight instructor during World War II.

Chesley Sullenberger—The pilot known for ditching a US Air A320 in the Hudson River off Manhattan, the "Miracle on the Hudson" that saved 155 lives, graduated from the Air Force Academy and flew F-4s in the late 1970s.

CHAPTER 10
MOVIES ABOUT THE AIR FORCE

"Like nothing you ever saw before... outside of a dream."
- Eugene Kern, narrator of the film *Memphis Belle*, 1943

Many films feature Airmen. These are some favorites.

Wings (1927). Directed by William Wellman. Richard Arlen, Buddy Rogers, Clara Bow. Wellman was a pilot in World War I, and this movie reflects his respect for fliers. The aerial combat scenes are superb. Arlen and Rogers are rival pilots and rivals for Clara Bow's heart. The first "Best Picture" and perhaps the last great silent movie, it is an excellent film.

Air Force (1943). Directed by Howard Hawks. John Garfield, Gig Young, George Tobias. After Pearl Harbor, Hollywood swung into the war effort. The outcome was far from certain when *Air Force* was being produced. This film is action-packed, as a B-17 crew fights earnestly but improbably at Pearl Harbor, at Wake Island, in the Philippines, and across the Pacific, culminating in an air raid on Tokyo. As one critic said, it was overboard but thrilling and fascinating.

Memphis Belle (1944), *Report from the Aleutians* (1943). These two documentaries feature the Army Air Forces in combat over Europe during the bloodiest portion of the bombing campaign against Germany, and in that Alaskan island chain, where weather was more lethal than the enemy, and boredom ruled. Directed by Hollywood greats William Wyler (*Belle*) and John Huston (*Aleutians*), these are well-made and remain interesting today. In 2018, documentarian Erik Nelson used 90 hours of outtakes from *Memphis Belle*, found in the National Archives, to create *The Cold Blue*, with narration and commentary from surviving World War II veterans. It is a welcome addition to our visual history of the Eighth Air Force in World War II.

Combat America (1943), produced by Clark Gable. *Winning Your Wings* (1942), directed by John Huston. Clark Gable flew combat missions while making *Combat America*. It was intended to recruit gunners, chronicle their work and reassure the folks back home that their boys were doing a good job. Gable banters with Airmen; they go to church and respectfully visit historic places in England. At last, we see a couple of gunners walking across a flightline, as the Air Force Song rises. Gable speaks the movie's closing lines: "Enemy of America, look at these men. They're not gonna lose, brother!" Syrupy? Yes. Rousing and patriotic? Yes. America's determination is clear. And so it is in the second film, *Winning Your Wings*. Accompanied by the Air Force Song, James Stewart taxies up in a trainer plane, hops out and, with a boyish grin, addresses the audience. America is building 100,000 planes and "somebody's got to fly them. A lot of somebodies." He lays out the path: leaving home, what trainees do, the money they were paid and the effect those silver wings had on the girls. It is hokey in a wholesome way, but American boys sitting in theaters were inspired to join after watching this.

Thirty Seconds Over Tokyo (1944). Directed by Mervyn LeRoy. Van Johnson, Spencer Tracy. Made during the war, this movie is an exciting adventure. Based on pilot Capt. Ted Lawson's book of the same title, it details the Doolittle Raid and the crewmen's experiences in China.

12 O'Clock High (1949). Directed by Henry King. Gregory Peck, Dean Jagger, Hugh Marlowe, Gary Merrill. The definitive study of the pressure of command in a war where your men are dying every day. The commander leads his men, they fight heroically and everyone pays a high price.

Strategic Air Command (1955). Directed by Anthony Mann. James Stewart, June Allyson, Frank Lovejoy. A baseball player is recalled to duty in the Strategic Air Command (SAC) to fly bombers as the Cold War heats up. The importance of SAC and its deterrent mission is highlighted.

Battle Hymn (1957). Directed by Douglas Sirk. Rock Hudson, Anna Kashfi, Dan Duryea, Don DeFore. Dean Hess was a minister who was a fighter pilot in World War II and Korea. He rescued hundreds of Korean children and established orphanages for war orphans. The movie dramatizes Hess's efforts but takes liberties with facts (as so many historical films do).

The Hunters (1958). Directed by Dick Powell. Robert Mitchum, Robert Wagner. This adaptation of James Salter's novel becomes a love triangle tale, diluting the drama of air combat and the men who range from an "it's my job" attitude to one that clearly revels in it as a blood sport. The aerial scenes featuring the F-86 Sabre are exciting.

Lafayette Escadrille (1958). Directed by William Wellman. Tab Hunter, Etchika Choureau, Marcel Dalio, David Janssen, Clint Eastwood. *Flyboys* (2006). Directed by Tony Bill. James Franco, Jean Reno, Jennifer Decker, David Ellison, Abdul Salis. Those Americans who volunteered to fly with the Lafayette Escadrille have not fared well on screen. Reviews of *Flyboys* were so-so, and the movie was a box office bomb. Wellman wanted *Lafayette Escadrille* to be his tribute to his flying comrades in the Great War, but he was disappointed in how it turned out. Reviewers panned the acting and called the story mediocre. Studio head Jack Warner insisted on script changes. Probably most hurtful to Wellman, even the surviving members of the Lafayette Escadrille panned the film. He never directed another picture.

A Gathering of Eagles (1963). Directed by Delbert Mann. Rock Hudson, Rod Taylor, Mary Peach, Barry Sullivan. In SAC's heyday, a new wing commander arrives after a failed inspection. He is brutally effective, even at the cost of personal relationships. An excellent portrayal of the pressures of life in SAC.

Dr. Strangelove, or How I Learned to Stop Worrying and Love the Bomb (1963). Directed by Stanley Kubrick. Peter Sellers, George C. Scott, Sterling Hayden. During the Cold War, fear of an accidental nuclear conflict was widespread. An insane commander launches the bombers and somehow, this scenario is played for laughs. *Dr. Strangelove* was popular among SAC crews for its mockery of Cold War thinking—a briefing book labeled "World Targets in Megadeaths," preposterous generals, and the absurdity of the President shouting, "Gentlemen, you can't fight in here, this is the War Room!"

Fail-Safe (1964). Directed by Sidney Lumet. Henry Fonda, Dan O'Herlihy, Walter Matthau, Edward Binns, Larry Hagman. This Cold War thriller examines the inadvertent launch of nuclear-armed bombers and the efforts to recover from the situation. The title refers to the geographic point at which bombers would orbit and await further orders. This is a dark film, as the President makes a horrific decision to avert an all-out nuclear war.

Catch-22 (1970). Directed by Mike Nichols. Alan Arkin, Bob Balaban, Martin Balsam, Richard Benjamin, Buck Henry, Bob Newhart, Paula Prentiss, Martin Sheen, Jon Voight, Orson Welles. A B-25 bombardier wants to escape the war by claiming he's insane. Alas, if you want to get out of a war, you're not insane. It's a parody of bureaucracy and the military, in many ways more a reflection of 1970 than World War II.

The Right Stuff (1983). Directed by Philip Kaufman. Sam Shepard, Ed Harris, Scott Glenn, Dennis Quaid, Fred Ward. The film covers Chuck Yeager's breaking of the sound barrier, 1950s test pilot culture and the early days of the Mercury program. It is a heroic, funny and noble story.

Good Morning Vietnam (1987). Directed by Barry Levinson. Robin Williams, Forrest Whitaker, Bruno Kirby. Adrian Cronauer was an American Forces Network disc jockey in Saigon during the early years of the Vietnam War. Robin Williams amps up a movie that is both hilarious and haunting. Cronauer sees the pain of the Vietnamese people and the brave young Americans far from home. Robin Williams did several USO tours after making this film, and his irreverent comedy brought laughs to deployed troops.

Bat-21 (1988). Directed by Peter Markle. Gene Hackman, Danny Glover. The story of Lt. Col. Gene Hambleton is presented here. Shot down on an electronic jamming mission, he is thrust into a survival situation. An SAC bomber navigator with experience in nuclear missile operations, Hambleton would have been quite a prize for the enemy. A complex rescue mission over several days finally plucks him from the jungle and the enemy. Gene Hackman and Danny Glover are excellent.

Memphis Belle (1990). Directed by Michael Caton-Jones. Matthew Modine, D.B. Sweeney, Billy Zane, Harry Connick Jr., Eric Stoltz. A fictional story about the final mission of the Memphis Belle in 1943. One Memphis Belle crewman said that everything depicted happened to some B-17 crew. It's not a great movie, but it's a good depiction of the bomber war in Europe.

By Dawn's Early Light (1990). Directed by Jack Sholder. Powers Boothe, Rebecca DeMornay, James Earl Jones, Martin Landau. As an accidental nuclear war breaks out, US flight crews and political leaders scramble to do what they think best. There is melodrama between bomber pilots Boothe and DeMornay. But it's the tension as US leaders try to figure out who is really in charge and what they should do that makes for a suspenseful film. James Earl Jones apparently had a good (fictional) Air Force career: He played a lieutenant in *Dr. Strangelove* and a general in this one.

Red Tails (2012). Directed by Anthony Hemingway. Terrence Howard, Cuba Gooding Jr., Nate Parker, David Oyelowo, Tristan Wilds, Ne-Yo. *The Tuskegee Airmen* (1995). Directed by Robert Markowitz. Laurence Fishburne, Cuba Gooding Jr., John Lithgow, Malcolm-Jamal Warner. *Red Tails* is highly entertaining. The characters are likable, and the aerial scenes are exciting. *The Tuskegee Airmen* was more straightforward in describing the challenges faced by black Airmen in overcoming prejudice and getting into the fight. It also makes clear Benjamin O. Davis's role in building this outfit.

Unbroken (2014). Directed by Angelina Jolie. Jack O'Connell, Domhnall Gleeson, Garrett Hedlund, Miyavi. The harrowing story of Louis Zamperini, Olympian turned bombardier, whose plane went down in the Pacific in 1943. After 47 days in a life raft, Zamperini and another American were captured and lived through two years of hell in Japanese prisoner-of-war camps. The book *"Unbroken"* by Laura Hillenbrand is a more comprehensive story, including Zamperini's postwar religious conversion that brought him peace after his horrifying experiences.

The Last Full Measure (2019). Directed by Todd Robinson. Sebastian Stan, Christopher Plummer, William Hurt, Ed Harris, Samuel L. Jackson, Jeremy Irvine. A solid account of Airman First Class William Pitsenbarger's valor in Vietnam, and the struggle by the soldiers he saved to get him the Medal of Honor.

Masters of the Air (2024). TV series, several directors. Austin Butler, Callum Turner, Anthony Boyle. This miniseries tells stories of the 100th Bomb Group during World War II. It uses several real people as characters and shows the dangers faced by fliers and the relationships among them. It is exciting, and many of the design details are well done, but the reasons for strategic bombing are unclear and historical facts are jumbled. It's not a documentary, but it is worth watching.

One place where the Army Air Forces and Hollywood met was in the making of films for training and propaganda. During World War II, the First Motion Picture Unit was tasked to make these films. These Airmen spent much of the war in Hollywood. The team included Ronald Reagan, Van Heflin, William Holden, Karl Malden, DeForest Kelley, Clayton Moore and George Reeves. We have a president, three Academy Award winners, a *Star Trek* doctor, the Lone Ranger and Superman, all in the same outfit. Not bad.

JIM PFAFF

CHAPTER 11
FICTIONAL AIRMEN

"Air power is the greatest force for peace and freedom the world has ever known."
- General Hap Arnold

Major Tony Nelson: In the 1960s TV series *I Dream of Jeanie*, astronaut Tony Nelson (Larry Hagman) finds a bottle containing a beautiful genie (Barbara Eden). He takes the bottle and the genie home to Cocoa Beach, Florida, where he's based with NASA. Comedy and romance follow.

Colonel Rainer Sarnac: In the 1984 TV series *Call to Glory*, Craig T. Nelson plays a Kennedy-era reconnaissance wing commander. The U-2 co-stars in this short-lived but compelling series. The characters are affected by events in the world in a way that military families can well understand. Heavily promoted during the 1984 Summer Olympics, the show was patriotic and showed the pride Airmen feel in service and the strains placed on military family members.

Brig. Gen. Frank Savage and Col. Keith Davenport: *12 O'Clock High* is the seminal Air Force leadership film. Davenport (Gary Merrill) commands a B-17 bomb group that is being decimated, as the 8th Air Force was in 1942. Casualties were high, results were low and hope was almost nonexistent. Davenport feels the losses keenly and tries to protect his men and build them up, even in the face of failure. Enter General Savage (Gregory Peck), who takes command with brutal honesty: "Stop making plans. Forget about going home. Consider yourselves already dead." He turns the unit around, and they become outstanding before he, too, collapses under the weight of command. Many World War II bomber veterans said this is the most realistic picture of what it was like to fly in the "Mighty Eighth."

Colonel Robert Hogan and his team: *Hogan's Heroes* managed to make a comedy of the dreary life of prisoners of war in Nazi Germany. The Allied POWs, led by the jovial Colonel Hogan (played by Bob Crane) engage in sabotage, conduct commando raids and occasionally slip into a local gasthaus for a beer. Hogan is charismatic, a smooth talker who spars with the German commandant and romances the secretary. Staff Sergeant Kincheloe (Ivan Dixon) is a radio operator who maintains an incredible communications network, spying on the commandant's office and keeping in touch with Allied headquarters in England. The team's demolition man is Technical Sergeant Carter (Larry Hovis), a goofy explosives expert. If you can accept the ridiculous premise, the show is funny.

General Buck Turgidson: *Dr. Strangelove* is a bizarre and funny satire of Cold War policies. General Turgidson (George C. Scott) is the Air Force Chief of Staff, an enthusiastic officer, proud of his service, of what it can do, and ready to fight the Communists, even though he accepts that US losses would be high: "I'm not saying we wouldn't get our hair mussed. But I do say no more than ten to twenty million killed, tops. Depending on the breaks."

Brig. Gen. Jack D. Ripper: In *Dr. Strangelove*, Ripper (Sterling Hayden) is the commander who has decided to start the decisive war against Soviet Communism to maintain the purity of our bodily fluids. Ripper is a steely-eyed warrior who believes that "war is too important to be left to politicians." He is single-minded in his devotion to the cause. That said, he does not avoid women, although he denies them his essence.

Also in *Dr. Strangelove*, Maj. T.J. Kong (Slim Pickens) and Lt. Lothar Zogg (James Earl Jones) are part of a B-52 crew flying General Ripper's mission. Maj. Kong is a country boy, ready for "nuclear combat toe-to-toe with the Russkies," enthusiastically going over survival kits and attack plans. Lt. Zogg is the Radar Navigator, Jones's first movie role. His character has no memorable lines, but it's a highlight when he pops up through the hatch into the cockpit and you hear that unforgettable voice.

Gen. James Mattoon Scott: *Seven Days in May* is a story about an attempted military coup. Burt Lancaster plays General Scott, a charismatic and much-decorated officer opposed to a treaty the president has signed with the Soviet Union. Despite all his combat decorations, this character is no hero. Made in 1963, it is still thought-provoking.

General Beringer: *War Games* was one of many anti-nuke movies produced in the early 1980s. A teenage hacker (Matthew Broderick) looking for computer games—in the days of dial-up connections and green text on black screens—has inadvertently gotten into a defense warning system, causing indications that the US is about to be attacked by the Soviet Union. He ends up at NORAD, where General Beringer is in charge. The general is fed up with the civilian whiz kids and their gizmos, even before this kid arrives. Barry Corbin, as General Beringer, bears a stocky, dark-haired resemblance to Gen. Curtis LeMay, and the hard-nosed general has not only a grim determination to face the situation but more common sense than the civilian brainiacs.

Col. Jim Caldwell (Rock Hudson) and Col. Hollis Farr (Rod Taylor): *A Gathering of Eagles* was filmed at the peak of SAC's influence. The depiction of a SAC wing undergoing a high-pressure, no-notice Operational Readiness Inspection (ORI) is realistic. Colonel Caldwell is the fast-moving officer on the climb, taking over a faltering wing. Colonel Farr is the holdover vice commander, a Keith Davenport to Caldwell's Frank Savage. As a drama, it's so-so. As a story of the demanding world of the Strategic Air Command in the Cold War, it is an excellent record of the times.

Lt. Col. Dutch Holland: In *Strategic Air Command*, James Stewart plays a major league baseball player recalled to active duty to fly B-36s. He flew B-29s in World War II and he's not happy about giving up his spot with the St. Louis Cardinals to go back into service. Of course, he does the right thing, even though it wrecks his baseball aspirations. It's a Cold War story of the price paid by those who weren't fighting a dramatic hot war but doing the hard labor of maintaining the peace. There is some beautiful footage of the B-36 in flight. On the ground, the B-36 was an ungainly bird. In flight, she was a graceful winged creature.

Capt. John Yossarian: *Catch-22* is about a B-25 unit in Italy in World War II. Yossarian has flown a lot of missions, is tired of the war and wants out, feigning insanity. However, the doctor tells him that anyone sane enough to want out of the war is mentally stable enough to keep flying and fighting. "Catch-22" has entered the English language as a term for an impossible dilemma. B-52 crewmen will tell you most "BUFFs" had graffiti scrawled somewhere proclaiming, "Yossarian was right!"

JIM PFAFF

CHAPTER 12
RANKS, INSIGNIA, UNIFORMS

"A career in the Air Force has inherent purpose and meaning unmatched in the civilian world. To me, it feels like the most noble thing to do with my life."
– Gen. Duke Richardson

As the Air Force gained independence, the question of what its uniforms should look like was important. Some wanted to keep traditional Army insignia and accoutrements. Many felt that insignia and accessories should be limited, and that all personnel should wear a uniform with only necessary insignia. A precedent: Marines wear few badges and accoutrements.

When the US Air Force was born in 1947, its members still wore Army-green uniforms. Air Force officers retained Army rank and insignia; enlisted rank and insignia evolved. Airmen wore polished black leather accessories instead of the Army's brown shoes and belts. Air Force blue uniforms were introduced in 1949, but olive-drab uniforms were still in use until 1952. The new USAF service dress was similar to the Army's, featuring four buttons and four pockets. Brass was replaced with oxidized silver insignia—no polishing needed—and no unit insignia was worn on the dress uniform. There was even an Ike jacket till 1964.[18]

Khaki was an Army tradition and summer service dress khaki uniforms lasted in the Air Force until about 1965, though men wore khaki short-sleeve shirts and trousers, known as 1505s, until 1978. Women wore light blue summer uniforms. The familiar light blue shirt over dark blue trousers came into use in the 1970s. Until the mid-1980s, an optional dark blue shirt was worn with the blue trousers. A noticeable change from Army practice was awarding a ribbon for time in service, instead of service stripes on the lower sleeve, and a marksmanship ribbon rather than a badge.

The biggest change came in 1993, when CSAF Gen. Merrill McPeak introduced

---

[18] Gen. Ira Eaker's aide, James Parton, said in his biography of Eaker, that Eisenhower admired Eaker's waist-length jacket, which he wore when flying. The waist-length jacket became closely associated with Eisenhower.

a new three-button service coat with no epaulets, and officers wore stripes on the lower sleeves denoting rank, Navy style (in silver braid rather than gold). Even the Air Force Seal on the coat buttons was changed to the "Hap Arnold" wing and star emblem. It had one fake pocket on the chest and two flaps without pockets, which looked useful but failed to be so. This uniform was as popular as acne. Even senior officers muttered that these uniforms made them look like airline pilots. The next CSAF brought back officers' epaulets and metal shoulder insignia at the speed of heat. (That's very fast: how long does it take to pull your hand away from a hot stove?)

US Air Force utility uniforms were called "the pickle suit." It was basically the Army's olive-drab green fatigues, with a blue belt, different name tapes, and US AIR FORCE markings. Airmen had white letters on a bold ultramarine-blue background. Rank insignia was brightly colored, with white stripes trimmed in blue for enlisted personnel, white for silver rank insignia and bright yellow for gold-colored rank insignia. These changed to subdued blue letters on a dark green background around 1980, and the rank became subdued, with dark blue stripes on a green background; black insignia represented silver officer ranks, while brown replaced gold. In the 1980s, the Battle Dress Uniform (BDU) became the standard, a popular woodland camouflage design. Desert uniforms appeared during Desert Shield in 1990, first in the "chocolate chip" pattern (dark flecks on a beige/brown uniform), and thereafter in a plain Desert Camouflage Uniform (DCU) in tan with blotches of pale green and dark brown. The BDU stayed in use in most of the world until 2011, but the DCU was worn in desert climates, primarily the Centcom Area of Responsibility (AOR).

The Airman Battle Uniform (ABU) replaced the BDU and DCU as a fatigue uniform. It was introduced in 2007 and became mandatory in late 2011. This uniform was not flame-resistant, and in 2010 (yes, before it was entirely fielded, it was obsolete), its replacement began appearing for Airmen whose duties required this protection. The replacement, the Army's Operational Camouflage Pattern (OCP), began to see wide use by 2018. The never-popular ABU was retired in 2021.

BERETS—Some Airmen wear berets. The beret's color identifies qualification: Black—Tactical Air Control Party (TACP); Dark blue—Security Forces; Maroon—Pararescuemen and Combat Rescue officers; Scarlet—Combat Controllers and Special Tactics officers; Gray—Special Reconnaissance and Weather Parachutist qualified Airmen; Sage green—Survival, Evasion, Resistance and Escape (SERE) specialists.

"FARTS AND DARTS"—The clouds and lightning bolts embroidery found on field grade and general officers' cap visors. The CSAF (and an Air Force CJCS) wears a hat with distinctive clouds and thunderbolts on the bill and around the cap itself.

JET NOISE IS THE SOUND OF FREEDOM

RANKS—Air Force enlisted ranks evolved as the service became less like the Army. Initially, Airmen were indistinguishable from their soldier counterparts: Private, Private First Class, Corporal, Sergeant, Staff Sergeant, Technical Sergeant, Master Sergeant. There were chevrons (pointing up) and rockers (the lower stripes). A First Sergeant would have a diamond between the rockers and chevrons. During World War II, the vast number of technical specialist personnel drove the creation of grades called technician–third, fourth, and fifth grades, corresponding to corporal, sergeant and staff sergeant, respectively. Technicians often were engine mechanics, electricians, radio or radar repairers and other skilled jobs. They wore a "T" below the chevrons.

Gen. Hoyt Vandenberg approved a new insignia design in 1948 with inverted stripes resembling wings. The rank titles initially remained the same as in the Army. In 1950, Vandenberg directed that people in the Air Force would no longer be called soldiers, but Airmen. The junior enlisted rank structure was revamped in 1952. Private (E-1) became Basic Airman, Private First Class (E-2) was now Airman Third Class, Corporals (E-3) were Airman Second Class, and Sergeants (E-4) became Airman First Class. The NCO ranks were Staff Sergeant (E-5), Technical Sergeant (E-6), and Master Sergeant (E-7).

In 1958, Congress approved two new senior enlisted ranks, pay grades E-8 and E-9. The Air Force chose to call the new grades Senior Master Sergeant and Chief Master Sergeant, with one chevron on top of the stripes for Senior Master Sergeants and two on top for Chief Master Sergeants. In 1967, the junior enlisted ranks were renamed Airman Basic (E-1), Airman (E-2), Airman First Class (E-3), and Sergeant (E-4). The three Airman ranks would wear a blue star in the center of the stripes; the NCOs would wear a silver one.

The E-4 rank was split in 1976, as newly promoted E-4's became Senior Airmen. These were not NCOs, but they were eligible to become Sergeants (E-4) upon meeting the qualifications. CSAF McPeak eliminated the E-4 sergeant rank in 1991 and returned all insignia stars to silver. The last "buck" sergeant left the USAF in 1998. Also in the early 1990s, Master Sergeants got a chevron at the top, giving Senior Master Sergeants two chevrons and Chiefs three. This change recognized the master sergeant as a senior NCO, akin to the Navy Chief Petty Officer or Marine Gunnery Sergeant. These ranks and insignia have remained the same since the 1990s.

Today, there are nine Enlisted Pay Grades, with E-1 being the entry point for most Airmen upon joining the military. Airman Basic (AB) is the most junior rank and the lowest paid. It's said that AB stands for "Always Broke."

There are ways to start a little higher, with one or two stripes. These include being an Eagle Scout, completing a JROTC course, or having a number of college credits. As you advance, both your pay and rank increase.

E-1 Airman Basic (AB)—no stripes. Sometimes called "Slick Sleeve" because there are no stripes on the sleeve.

E-2 Airman (Amn)—one stripe.

E-3 Airman First Class (A1C)—two stripes.

E-4 Senior Airman (SrA)—three stripes.

E-5 Staff Sergeant (SSgt)—four stripes.

E-6 Technical Sergeant (TSgt)—five stripes.

E-7 Master Sergeant (MSgt)—five stripes and one chevron.

E-8 Senior Master Sergeant (SMSgt)—five stripes and two chevrons.

E-9 Chief Master Sergeant (CMSgt)—five stripes and three chevrons. A Chief Master Sergeant may be addressed as "Chief." Similarly, a Senior Master Sergeant may be called "Senior." To become a Chief Master Sergeant is a tremendous achievement. The title, respect and prestige are heartfelt. In contrast, for a Chief Master Sergeant to be called an "E-9" is an insult. Being called an E-9 means one may wear the rank, but isn't worthy of the name "Chief."

Sergeants from E-5 through E-9 are Non-Commissioned Officers. NCOs are expert technicians, trainers and first-line supervisors. NCOs drive mission accomplishment, performance, and unit success. NCOs build an atmosphere of loyalty, provide mentorship and carry on the ideals of the Service.

There are still First Sergeants, with a diamond between the stripes and chevrons. There are Command Chiefs, the senior enlisted members in the unit, with a star between the stripes and chevrons. At the enlisted pinnacle is the Chief Master Sergeant of the Air Force (CMSAF). This insignia is elaborate, with a wreath around the star in the center of the stripes and the Coat of Arms of the United States and two stars between the stripes and chevrons. The CMSAF is the highest ranking active-duty enlisted member and an adviser to the CSAF. CMSAF provides guidance for the enlisted force and represents their interests in public forums, before Congress, and among senior military leadership. CMSAF is appointed by CSAF and advises on all issues regarding the welfare, readiness, morale and proper utilization of the enlisted force.

The most revered CMSAF recently has been "Enlisted Jesus," Kaleth O. Wright. He was beloved for eliminating Enlisted Performance Reports for E1-E3, championing the Operational Camouflage Pattern and ditching the Airman Battle Uniform, and his advocacy for the enlisted force. He was one of the few leaders who could fill a base theater (usually the largest auditorium on the installation) with people eager to hear him speak.

Warrant Officers rank between the most senior enlisted (E-9) and the most junior commissioned (O-1) ranks. They receive the courtesies of a commissioned officer. There are five warrant officer grades. During World War II, it was common to see pilot, navigator and bombardier "Flight Officers," assuming they would fly but not be in leadership roles. The Air Force had warrant officers through the 1950s, but when the enlisted ranks were expanded to include E-8 and E-9, USAF leaders believed there was no longer a need for a warrant officer. (The other services disagreed.) The Air Force stopped creating warrant officers in 1959, and the last on active duty, CWO4 James H. Long, retired in 1980. Today, the Air Force defines a warrant officer as "a technical specialist with supervisory ability." Subject matter experts with the authority of officers, they are advisers and problem solvers for commanders. In 2024, the Air Force revived the warrant ranks for cyber and information management specialists and some positions in the RPA community.

Air Force commissioned officer ranks are identical to those of the Army: same titles and insignia. The only difference is at the five-star level. The Army has a General of the Army; the Air Force has a General of the Air Force. The Army's five-stars were George C. Marshall, Douglas MacArthur, Dwight D. Eisenhower and Henry H. Arnold. Later, Omar N. Bradley was awarded a fifth star. General Arnold left active duty in 1946, but five stars are never considered retired. They receive full pay and allowances for life, so in 1949, Arnold was transferred into the Air Force. He remains the only five-star general in Air Force history.

There are ten commissioned officer ranks. Lieutenants and captains are company-grade officers, a term from the Army. Majors and colonels are field grade (again, an Army term). Then there are generals.

From most junior to most senior, with their rank insignia:

Second Lieutenant (2Lt)—one gold bar. Second lieutenants are sometimes called "butterbars" because their gold bar insignia resembles a stick of butter.

First Lieutenant—one silver bar. A familiar term for a lieutenant, sometimes used as a form of address, is LT (pronounced "EL-tee").

Captain—two silver bars. Captains' rank is occasionally called railroad tracks, because the bars are parallel, like train tracks.

Field grade officers are:

Major—gold oak leaf.

Lieutenant Colonel—silver oak leaf.

In flying squadrons and headquarters, there are often lieutenant colonels who are unlikely to be promoted. They are frequently reassigned to wing staff duties or positions where their expertise can be utilized. Their work areas may be called the "Silver Leaf Lounge," in recognition of the rank insignia. It's a parking area for people moving toward the end of their careers. A "phone colonel" is a lieutenant colonel who doesn't identify himself as such during a phone call, relying on the word "colonel" to lend his words weight. Only about half of the lieutenant colonels will be promoted to colonel, so it is significant.

Colonel—silver eagle. Colonels may be called a bird colonel or a "full bird" (to distinguish from lieutenant colonels). These terms come from the colonel's eagle rank insignia. A colonel may also be called an "O-6." It's not an insult like being called an "E-9". It's simply a reference to the colonel's pay grade. Colonels are often the highest-ranking officers on a base unless there is a higher headquarters. There, they are usually senior staff.

General officers all wear silver stars. They may be called GOs (pronounced "Gee-oh"). They are officers in pay grades O-7/8/9/10. They may be called a flag officer (more common in the Navy) because these ranks are honored with ceremonial flags denoting the number of stars they wear.

Brigadier General—one star.

Major General—two stars.

Lieutenant General—three stars.

General—four stars.

There are only a dozen or so four-stars on active-duty in the Air Force at any time. They command the highest echelons of the military. The CSAF is a four-star, and is the USAF's senior ranking officer, though there could be a USAF general as chairman or vice chairman of the Joint Chiefs of Staff. The chairman of the Joint Chiefs Of Staff (CJCS) is the president's adviser on military affairs. In August 1957, Gen. Nathan F. Twining became the first USAF officer to serve as chairman of the Joint Chiefs of Staff. He remained in that position until September 30, 1960.

Gen. George S. Brown was CJCS from July 1974 to June 1978. He was followed by Gen. David C. Jones, who served until June 1982. Gen. Richard Myers was the next USAF CJCS, from October 1, 2001, till September 30, 2005. Unlike the others, he had not been Chief of Staff of the Air Force (CSAF), but went from NORAD chief to vice chairman before his selection for the top job. The most recent CJCS wearing Air Force blue was Gen. Charles Q. Brown, who moved up from CSAF to the CJCS post in late 2023. Brown's successor, Gen. Dan Caine, is an Airman as well.

The highest possible rank is General of the Air Force—five stars, arranged in a circle. Only General Henry H. Arnold achieved this rank.

JET NOISE IS THE SOUND OF FREEDOM

AIR FORCE CHIEFS OF STAFF (CSAF)

If my goal were to be CSAF someday, I would go to the Air Force Academy and be a fighter pilot. For most officers, the source of commission doesn't matter. And while flying fighters is exciting and cool, flying "the heavies" (bombers, tankers, transports) is also an interesting and exciting job. Many other valuable and fulfilling roles are not flying jobs. Fewer than 1 percent of the lieutenants entering the Air Force will wear general's stars. Still, many rewarding and satisfying careers are completed as field-grade officers. Here are the Air Force chiefs of staff, their backgrounds, operational experience, and source of commission:

Kenneth S. Wilsbach—Fighter, Persian Gulf. AFROTC.

David W. Allvin—Airlift, Persian Gulf. USAFA.

Charles Q. Brown—Fighter, Persian Gulf. AFROTC.

David L. Goldfein—Fighter, Persian Gulf, Kosovo. USAFA.

Mark A. Welsh—Fighter, Persian Gulf, Vietnam. USAFA.

Norton A. Schwartz—Airlift & Special Ops, Persian Gulf. USAFA.

T. Michael Moseley—Fighter, Persian Gulf. AFROTC.

John P. Jumper—Fighter, Vietnam. AFROTC.

Michael E. Ryan—Fighter, Vietnam. USAFA.

Ronald R. Fogleman—Fighter, Vietnam. USAFA.

Merrill A. McPeak—Fighter, Vietnam. AFROTC.

Michael J. Dugan—Fighter, Vietnam. West Point.

Larry D. Welch—Fighter, Vietnam. Aviation cadet.

Charles A. Gabriel—Fighter, Korea, Vietnam. West Point.

Lew Allen—Bomber/Weapons Research, Cold War. West Point.

David C. Jones—Bomber, Korea, Cold War. Aviation cadet.

George S. Brown—Bomber, WWII, Korea, Cold War, Vietnam. West Point.

John D. Ryan—Bomber, WWII, Cold War, Vietnam. West Point.

John P. McConnell—Fighter, WWII, Cold War. West Point.

Curtis E. LeMay—Bomber, WWII, Cold War. ROTC.

Thomas D. White—Fighter/Intel, WWII. West Point.

Nathan F. Twining—Bomber, WWII. West Point.

Hoyt S. Vandenberg—Fighter/Attack, WWII. West Point.

Carl A. Spaatz—Fighter/Bomber, WWI, WWII. West Point.

AWARDS AND DECORATIONS

People notice the colorful ribbons on military uniforms, which recognize service and valor. An award is given for meeting some general criteria (service in a particular campaign or activity), while decorations recognize valor or achievements of personal merit.

The Medal of Honor is the highest award for bravery "above and beyond the call of duty." This means acts performed at great risk of life, actions that, had they not been done, no one would have faulted the individual. Of the 60 Airmen awarded the Medal of Honor, 35 did not survive the action.

The next award is the Air Force Cross for extraordinary heroism in combat. This was created in 1960, and about 202 have been awarded. Fifty awards were posthumous. Before there were Air Force-specific medals, Airmen received Army awards. The Army equivalent is the Distinguished Service Cross.

Several World War I fliers won multiple DSCs. Eddie Rickenbacker, with 26 kills, received seven DSCs. Douglas Campbell, an ace with six victories, won five DSCs, as did Frank Hunter, who scored nine victories. Reed Chambers, a squadron mate of Rickenbacker and Campbell, shot down seven Germans and was awarded four DSCs. Murray Guthrie, Ralph O'Neill and Glen Preston, a reconnaissance observer, each earned three of the prestigious honors.

After World War I, only two Airmen were awarded three Distinguished Service Crosses: William Whisner, an ace of two wars, and John C. Meyer. Both were fighter aces in the European Theater in World War II. Whisner received two DSCs in World War II, and one more in the Korean War. Meyer earned all three of his DSCs in seven months, May 1944-January 1945. Meyer ended World War II as a lieutenant colonel with 24 aerial victories. In the Korean War, he commanded the 4th Fighter Wing and shot down two more enemy planes.

There are many other decorations and awards today. Silver Stars, Bronze Stars, Distinguished Flying Crosses and Air Medals are awarded primarily for combat service. The Airman's Medal is given for non-combat heroism.[19] There are several levels of commendations, often duplicated by "joint service" awards. There are far more awards available for good service than for combat or valor. These are awarded at the commander's discretion, and failure to award commensurate decorations to deserving Airmen is a sign of poor leadership. There are many "I was there" awards—not just campaign medals, but overseas service (that can get you two ribbons!), service in a unit with a nuclear mission, Reserve duty, performing "special" duty, volunteer work, and a half dozen unit awards. Graduating from Basic Military Training earns

---

[19] An Airman I served with was awarded the Airman's Medal for saving people from drowning in a rip current off Guam. He was quite modest about it, but I was always impressed by the words "For Valor" inscribed on the medal.

a ribbon, as does serving four years. It's not unusual to see an Airman with more ribbons than Curtis LeMay. Such participation awards lessen their impact. Once, people outside the military were impressed by a "chest full of ribbons." Now, the impact of these enormous ribbon racks is diminished by their ubiquity.

A couple of terms used frequently in the Air Force: DFC, Distinguished Flying Cross, a significant award for heroism or achievement in flight. OAY, Outstanding Airman of the Year. The Air Force selects 12 Outstanding Airmen of the Year. Just 12 out of 270,000 enlisted Airmen is a prestigious award.

A few Airmen have received the Congressional Gold Medal. The Wright Brothers and Charles Lindbergh were so honored. Some groups have been recognized: the Tuskegee Airmen, Women's Air Service Pilots of World War II, American Fighter Aces, the Civil Air Patrol in World War II, and the Doolittle Raiders. Five individual Airmen have been so honored: Billy Mitchell, Buzz Aldrin, Michael Collins, Ira Eaker and Ronald Reagan. Chuck Yeager received a special congressional silver medal, "equivalent to a non-combat Medal of Honor."

JIM PFAFF

CHAPTER 13
AIR FORCE CULTURE AND LORE

*"There has never been a time when
I've been completely satisfied with myself."*
- Jimmy Doolittle

Even after Vietnam, America faced Soviet Communism. With its repression, police state brutality, and gulags, it was an evil empire. In those Cold War days, it seemed we would face the Communists for a long time, that the struggle against the Soviets would be like building a cathedral, the work of generations. The promises of the officers' oath: "to support and defend the Constitution of the United States against all enemies" seemed like the words consecrating a knight on a crusade.

So I wanted to go on a crusade, to fight for the right, defend freedom, and protect the weak. I wanted my life to count for something, to make a difference. I wanted to be able to look the veterans in the eye and say, "I did my part for our country." I was afraid that life would pass me by without having done some noble and lasting thing, and to have taken part in an adventure—that eternal dream of every boy.

Along the way, the Air Force culture and legends left their mark on me as well. This chapter offers a small sample of that legacy.

For many years, the Air Force was a garrison force on established bases. During the Vietnam War, Airmen were deeply involved in Southeast Asia, most serving there one year at a time. In 1983, I joined the Air Force, primed to deter a Soviet threat through readiness and strength in being. The post-Cold War USAF was expeditionary, ready to deploy frequently, answering the bell like a fire brigade. It seemed odd to me that a reduced force was tasked to deploy more regularly.

Life in the military can seem alien to civilians. Every military organization has much in common with the others, but has its own culture, ways and norms. Many Airmen use flight-related terms colloquially and understand air warfare as the Air Force's raison d'être.

A member of the Air Force is an Airman: If you're in the Army, you're a Soldier; if you're in the Navy, you're a Sailor; if you're in the Marines, you're a Marine; and if you're in the Air Force, you're an Airman. It's also part of the rank structure.

The Air Force's mission is to "fly, fight, and win... airpower anytime, anywhere." Its motto: "Aim High... Fly-Fight-Win."

Every Airman must understand and practice the AIR FORCE CORE VALUES: *Integrity, Service Before Self, and Excellence in All We Do.*

An Airman is expected to do what is right even when no one is looking.

An Airman is expected to place duty above personal desires.

An Airman is expected to perform duties at the best possible level.

No one following these ideals can go far wrong.

THE AIRMAN'S CREED is a statement of what serving in the Air Force means, a summary of values, guiding principles and a reminder of the seriousness and importance of the Air Force's mission for the United States:

I am an American Airman. I am a warrior. I have answered my nation's call.

I am an American Airman. My mission is to fly, fight and win. I am faithful to a proud heritage, a tradition of honor and a legacy of valor.

I am an American Airman. Guardian of freedom and justice, my nation's sword and shield, its sentry and avenger. I defend my country with my life.

I am an American Airman. Wingman, leader, warrior. I will never leave an Airman behind. I will never falter, and I will not fail.

The Air Force Song is a cheerful march that originated in a magazine contest. *Liberty Magazine* worked with Hap Arnold in 1938-39 to find a song for the Army Air Corps. Music teacher Robert Crawford submitted the winning composition, selected from more than 750 entries. Crawford was a pilot in the Air Transport Command in World War II, who returned to teaching and composing after the war.

Composer Samuel Barber, a World War II Airman, wrote a funeral march using the Air Force Song in a minor key. It is rarely played, but it is a somber and moving work.

The Air Force Memorial in Arlington, adjacent to the National Cemetery, was dedicated by President George W. Bush on October 14, 2006. This memorial honors the service and sacrifices of Airmen. The site is adjacent to Arlington Cemetery, near the Pentagon. Architect James Ingo Freed created a design with three stainless steel spires soaring into the sky, the tallest reaching a height of 270 feet. This evokes the majesty and power of flight and resembles the Thunderbirds' "bomb burst" maneuver. There is a bronze Honor Guard statue, and glass and granite walls with inspiring inscriptions about Air Force history, and others that recognize Medal of Honor recipients and fallen Airmen.

Joining the Air Force starts with meeting a recruiter. There will be a physical examination at a Military Entrance Processing Station (MEPS), where people are screened via medical testing and evaluation of their fitness for military service. There is a background check to ensure a recruit isn't a criminal, and the Armed Services Vocational Aptitude Battery (ASVAB), a multiple-choice test used to determine enlistment eligibility in the United States Armed Forces; it also determines the jobs for which an enlistee qualifies. A score of 50 is about average. Many Air Force jobs require a score of at least this level; you might be accepted with a lower score, but your job choices will be limited. Higher scores may open many more career fields.

Those entering the military incur an Active-Duty Service Commitment (ADSC). Active-duty enlistments typically last four to six years, depending on the length of training and bonuses offered, and may stipulate a term of Reserve duty following the completion of active duty. Enlisted members reenlist or separate from the Air Force as these commitments expire. Most officers have a four-year active-duty commitment, but pilots incur a 10-year commitment, and other rated officers, six years. Officers generally serve for an indefinite time after fulfilling their ADSC.

It is possible to join with a higher rank than Airman Basic. Air Force Junior ROTC (AFJROTC) is a high school citizenship and community service program for more than 90,000 students, taught by retired Air Force members at more than 800 high schools. It is not recruiting but teaches life skills, personal responsibility, character and self-discipline to develop citizens of character. Students who complete the JROTC program may enter the military at higher pay grades. It is also possible to join the Air Force with a higher rank by having sufficient college credit or being an Eagle Scout. Details on these programs are subject to change.

There are several ways to be an Airman: A person serving on active duty (AD) is in the military, 24/7. Active-duty personnel can be assigned to carry out the mission anytime, anywhere. Air Reserve Component (ARC) members are in the Air National Guard or Air Force Reserve; after initial training, they serve two days a month and two weeks of Annual Training (AT) each year, unless called to AD.

RESERVE COMPONENTS

In 1970, Defense Secretary Melvin Laird announced the Total Force Concept, making the Reserve and National Guard the initial sources for augmenting the active force. The Air Force and the Air Reserve Component (ARC: Air Force Reserve and Air National Guard) work well together. The result is that ARC forces take on a far greater role in normal USAF operations than other reserve components do with their parent services.

The Air National Guard (ANG) is part of the National Guard, which began as colonial militia; the Air Guard started during World War I. The ANG has a dual mission: It is a military reserve but also responds to the respective states during weather emergencies, disasters or unrest. In peacetime, state governors command the Air National Guard. In times of conflict or emergency, the president may activate the Guard.

The Air Force Reserve augments the active force when needed. There are a handful of full-time people who staff these ARC units. There are Active Guard Reserve (AGR) personnel in a few full-time military billets, and Air Reserve Technicians (ART), civil service employees who must be reservists in the unit where their civilian jobs exist. These members perform the day-to-day duties that keep the unit running between drill weekends. A flying wing with more than a thousand people assigned—and present on drill weekends or in case of mobilization—probably operates seven days a week, but may be staffed by only about 100 people on a non-drill day.

Civil Air Patrol (CAP) is not a military service but is an unpaid volunteer group. CAP was founded on December 1, 1941, to fly light aircraft patrols along US coasts to search for enemy submarines. In 1943, CAP became an auxiliary of the AAF. Although CAP is a civilian organization, it was awarded campaign credit for its antisubmarine missions and received the American Campaign Medal. Twenty-six of its members died in the line of duty. Today, its roles are in disaster relief, search and rescue, and aerospace education for youth.

LIFE IN THE AIR FORCE

Pay and benefits are good. In addition to a monthly paycheck, a member receives benefits including 30 days of paid vacation, free medical and dental care, tax-free food and housing allowances, tuition assistance, and low-cost insurance. The SRB (Selective Retention Bonus) can be a good thing. The Air Force pays reenlistment bonuses to those willing to serve—or extend service—in certain career fields. Enlisted aircrew, maintenance, cyber, medical and Special Operations AFSCs (Air Force Specialty Code) are often in high demand. Those who stay until retirement are eligible for a pension. Members may put aside money and receive matching funds, similar to a civilian 401(k) plan to save for the future, even if they do not make a career in the military. A person does not get rich in the military, but it can be a good life.

JET NOISE IS THE SOUND OF FREEDOM

For most people, entering the Air Force means Basic Military Training (BMT). It currently lasts about eight and a half weeks, but it's been longer or shorter over the years. It was conducted at Amarillo AFB, Texas, and Sampson AFB, New York, but for many years, Lackland AFB at San Antonio, Texas, has been the "Gateway to the Air Force." At BMT, trainees live, eat, work, exercise and learn together. For the first couple of days at Lackland, the newest trainees are described as a "rainbow flight." They have not yet received uniforms and are still wearing civilian clothing in a lot of colors, like a rainbow. But soon, uniforms are issued. Those who need eyeglasses will receive them, though the style is hideous enough to acquire the name BCD: Birth Control Devices. (Who's going to be interested in someone wearing these!)

They quickly learn the Reporting Statement: "Sir/Ma'am, (rank) (last name) reports as ordered." It stays with you forever and will be used at any time when responding to a superior, regardless of your rank. The trainees are led by Military Training Instructors (MTI), the Air Force version of the drill sergeant. MTIs must be leaders, with judgment, professionalism and an outstanding military image. They can yell and roast a trainee, shouting "What the piss!" when confronted with buffoonery, foolishness or just plain failure by trainees. Each trainee carries copies of AF Form 341, "Excellence/Discrepancy Report." An MTI or other authority may take one for a write-up and then file it as a demerit.

Form 341 can also recognize positive achievements, but Airmen say that's as elusive as Bigfoot: Someone saw it once, but nobody got a clear photo. The Snake Pit, where MTIs eat at BMT, is infamous. Trainees must walk by it to get to their tables. MTIs instantly discern any flaws in a trainee's uniform, appearance, shoes or demeanor. Trainees are quizzed on topics they have learned. It's stressful as trainees are grilled and corrected. There are legends of mythical Airmen who were never caught in the Snake Pit, but their names are lost in the mists of time.

As trainees develop into Airmen, furiously hollering MTIs may offer a "good job" or a nod, a subtle, confidence-building recognition. BMT is not the physically harsh regimen of the Marines or Army; rather, it involves constant exercise and running, and instills a sense of duty and self-motivation. Most days begin around 0500 (5:00 a.m.) and end at 2100 (9:00 p.m.). Other than saying there are three meals daily, describing the training schedule is impossible. Priorities within the Air Force change, and weather or unforeseen circumstances may shift things.

PACER FORGE (Primary Agile Combat Employment Range, Forward Operations Readiness Generation Exercise) is held during BMT to simulate some of the requirements of deployment: weapons handling, casualty care and other aspects of life in the field. This replaced BEAST (Basic Expeditionary Airman Skills Training) in 2022.

A persistent slander against the Air Force is that trainees can pull a Stress Card, and the mean ol' sergeant has to offer them a cookie or something. I think the famously salty Marine Chesty Puller made this one up. It has never existed. Maybe it sounds plausible in our therapeutic age. One thing that is part of BMT is "Ropes." Basic and Technical training students may wear colored shoulder cords, denoting selection for extra responsibility. Top Military Training Instructors are awarded a blue rope worn on the "Smokey the Bear"-style campaign hat.

Becoming an Airman is called becoming "blue," from the blue uniform and the many blue items accompanying it (hats, jackets, PT gear, etc.). The recruiting slogan "Cross into the blue" carries that theme. Airmen are occasionally called "bluesuiters." Indeed, later in a career, when Airmen attend training programs that renew their sense of being an Airman, it is called "re-bluing." That may come from the many uses of blue in the service or from re-bluing a firearm to keep it clean and usable.

After completing BMT, Airmen complete technical training in their career fields, primarily at five places: Goodfellow, Lackland, and Sheppard Air Force bases in Texas; Keesler AFB, Mississippi; and Vandenberg SFB, California. The training location depends on the Air Force Specialty Code (AFSC), which is how the Air Force identifies jobs within the service. (The Army and Marines call this a Military Occupational Specialty, or MOS, and the Navy calls it a rating.) The Airmen will have more freedom, but are still in training, supervised by Military Training Leaders (MTL). Some Airmen will learn their jobs in other locations, possibly even with another service. For example, Air Force Explosive Ordnance Disposal technicians attend a Navy school in Florida. Commissioned officers attend technical training courses at the same locations.

On-the-job training (OJT) is a local process, completing the technical training with hands-on training, to certify personnel in their jobs and award a skill level and job qualification.

The work life of an Airman depends on their assigned job. In personnel, finance, administration, acquisition, supply, or the many "office" jobs in the Air Force, the Airman usually works a schedule familiar to civilians—arrive around 7:00 a.m., leave around 5:00 p.m., etc. "Chair Force" is a term sometimes used to refer to the many desk jobs in the Air Force. Airmen roll their eyes and press on. This also gives rise to the joke that America has two armed services: The Army and the Navy; the Air Force is a corporation, and the Marines are a cult. A significant portion of the Air Force works a different schedule. Fliers and maintainers can be found at work around the clock. Missileers, missile maintenance and Security Forces never close their doors. And many others—command posts, firefighters and hospitals—work nights, weekends and holidays. This isn't to downplay anyone's contributions. It is simply acknowledging that there are differences.

Life in the Air Force revolves around the base. Most Airmen are stationed on a military base. It can be an Air Force Base (AFB), Space Force Base (SFB), Air National Guard Base (ANGB), or an Air Reserve Base (ARB). There are also Joint Bases (JB) and Joint Reserve Bases (JRB). Following the consolidation of military installations after the Cold War, many functions were combined, resulting in the creation of new bases, such as JBSA (Joint Base San Antonio), Joint Base Lewis-McChord and Joint Base Andrews. NAS JRB Carswell at Fort Worth, Texas, is a good example. For fifty years, this was Carswell AFB. In 1994, the Air Force left, and the Navy moved its operations from Dallas to Carswell, changing the base to a Naval Air Station (NAS). Then, the Navy was joined by the Air Force Reserve, Navy Reserve, Marine Corps Reserve, Army Reserve and Texas ANG units.

Where are Air Force bases? The Navy and Marines are usually on the coasts. The Army tends to be out in rural areas with room to maneuver. But Air Force bases are found all over. During the Cold War, SAC placed many of its bomber and missile installations deep in the heartland and as far north as possible to make them less accessible to attackers and closer to targets in the USSR. Fighter bases tend to be in warmer, sunnier climates, conducive to flying. Airlift bases are often on the coasts, enabling global reach. Training bases are generally in areas of favorable weather. The Air Force Reserve uses several former active base locations, often near large cities due to proximity to airline pilot domiciles and recruiting opportunities. The Air National Guard has locations in all 50 states, the District of Columbia, Puerto Rico, the Virgin Islands and Guam.

Each service has groups that advocate for its people and mission. The Air and Space Forces Association (AFA) is a lobbying group advocating for air and space power-related issues. The Air Force Sergeants Association (AFSA) lobbies for quality-of-life initiatives from the perspective of the enlisted force. Groups such as the Order of Daedalians[20], the Airlift-Tanker Association, the Air Commando Association, the Association of Air Force Missileers and many others keep camaraderie alive and honor those in their areas of expertise.

FUN AND GAMES

Celebrations are a great part of military life. One event is the arrival of a new commander. As a welcome, there is often a No-Notice Hospitality Check. This involves just showing up at the boss's house to eat, drink and socialize. It may sound rude, but in some (particularly overseas) assignments, keeping beverages and snacks on hand for such unannounced visits is not unusual. It's a morale-builder and a lot of fun. It might even begin with a "roof stomp," as Airmen walk around on the roof, demanding entertainment, food and

---

[20] Many of these groups are self-explanatory but Daedalians is an association of military aviators. The name is taken from the mythical Daedalus who made wings to escape imprisonment by flying away.

drink. A good-natured response by the host goes far toward acceptance and loyalty among the unit's people, and smart Airmen will coordinate this event with the boss's spouse. The roof stomp is a form of the Spirit Mission, a good-natured prank to show unit or team pride. It's usually harmless, like swiping a mascot or some token from another unit, or some other friendly rivalry.

Air Force units often have a summertime cookout to celebrate the warm weather, a successful mission, or just because it's Friday. Hosted by the commander and unit leaders, it's called a "Burger Burn," an apt name. Many things are fun: cookouts, happy hours, Christmas parties. But when you're directed to be there, it might not seem like fun, but more like another work event. This is Mandatory Fun, and just as in civilian life, a misstep can be career-changing. Nobody's taking attendance, but if you want to move up, you participate. Enjoy in moderation.

CRUD is a game played in Air Force clubs. Using a billiards table with a cue ball and a striped ball, the object is to "shoot" (by hand, no cue sticks) the striped ball into the table pockets with the cue ball despite the other team's efforts to distract or block the shooter. Teams may have a few or many players. The game is fast, it sometimes gets rough (body checks are common), and alcohol may be involved. It's every bit as fun and ridiculous as it sounds. Oh, did I mention you only point at opponents with your elbow? Ridiculous.

A unique Air Force celebration is Mustache March, when Airmen grow mustaches in honor of Robin Olds. When Olds commanded a fighter wing during the Vietnam War, his men grew "bulletproof" non-regulation mustaches. His was a magnificent handlebar, and he famously wore it when he met the Air Force Chief of Staff. Not surprisingly, he was ordered to get rid of it. Olds said, "The mustache became my silent last word in the verbal battles ... with higher headquarters on rules, targets and fighting the war." Olds shaved when ordered to, and on April 1, most Airmen are clean-shaven again.

While the military has rules on grooming, mustaches have long been seen among Airmen. Chief Signal Officer Brig. Gen. James Allen had a handlebar mustache in 1907 when, from the commander-in-chief (President Theodore Roosevelt) on, mustaches were popular. Many Airmen have worn facial hair: Carl Spaatz wore a mustache, as did Generals Mason Patrick, Oscar Westover, Gordon Saville, George Stratemeyer, Muir Fairchild, Leon Johnson, William Kepner, Larry Kuter and Frank Hunter. Robin Olds and his deputy commanders, Vermont Garrison and Chappie James, were mustachioed. World War II fighter aces Don Gentile, George Preddy and John Godfrey wore pencil-thin mustaches like fellow Airman Clark Gable. Lt. Gen. Dan James wore a mustache as did his father, Chappie.

Hail and farewell is a tradition. There are always people coming to or departing from a military organization, so once a month or so, there is a social event to welcome and say goodbye to those moving. The Fini Flight is a flier's last flight before retiring. After that last landing, the honoree celebrates with friends and family. Retirements are celebrations. Although leaving a beloved profession can be tough, moving on at an age where there is still plenty of opportunity is a great perk of a military career.

Airmen take care of their own: The Air Force Assistance Fund (AFAF) is an umbrella of charities that supports Airmen, retirees and dependents with emergency funds, education loans and retirement homes for widows or widowers of Air Force members. Another way Airmen help one another is the Airman's Attic, a "secondhand" store on base that offers donated furniture, household items, small appliances and clothing (including uniforms) at no cost to junior personnel and their families.

Each base is a unique community. Most jobs that you can find in civilian life are found in the Air Force, in addition to the strictly military tasks that people think of first. Most Airmen do not fly, but perform support roles necessary to support the mission.

There is a First Term Enlisted Center (FTEC), formerly known as FTAC (First Term Airmen's Center), to help Airmen transition from technical school to their first duty station. There are dependents, family members of the military person. The kids are lovingly called "Brats." A system called Defense Eligibility Enrollment Reporting System (DEERS) registers members and families for medical benefits, ID cards, GI Bill benefits, etc. There are schools, a hospital, and almost everything you might find in small-town America. The Army and Air Force Exchange Service (AAFES, pronounced "A-Fees") runs stores called the Base Exchange (BX), the department store on base, the Class Six store (where beer, wine & liquor are sold), and the Shoppette (convenience store). There is a Commissary (supermarket) and housing is available on most bases, ranging from okay to very nice.

The Force Support Squadron (FSS) provides many of these functions. FSS includes personnel, dining and lodging, mortuary and laundry, recreational and athletic facilities, family support activities, educational activities, child care and recreation facilities, through MWR (Morale, Welfare, and Recreation). These support Airmen, their families and military retirees, and are partially funded by the "profits" earned by AAFES. MWR provides gyms, pools, tennis courts, disc golf, golf courses, skeet and trap ranges, bowling centers, campgrounds and other outdoor activities, restaurants and clubs. Through the ITT (Information, Tickets, and Travel) office, members can obtain reduced admission to events and activities (sports, movies, concerts, shows, attractions such as water parks, theme parks). These may vary from base to base: Eielson AFB in Alaska has a ski slope, snowboarding and cross-country skiing; Tyndall AFB in Florida has beautiful white sand beaches and a marina.

Overseas bases may offer superb amenities. Overseas duty tours may be long tours—typically three or four years, accompanied by the member's family; and short tours—usually one year, often in austere locations, unaccompanied by family. A tour of duty in Japan will offer different opportunities than one in Italy. But the chance to spend time in a foreign land is priceless.

The US Postal Service works with American military personnel overseas to ensure their mail is delivered promptly. Mail is addressed to APO AE (Army/Air Force Post Office Europe) or APO AP (Army/Air Force Post Office Pacific), with a numerical designator for specific locations. This enables military people to use US postage and follow US postal rules. Overseas, military personnel are supported by the American Forces Network (AFN) that provides television and radio to US forces. Until 2009, it was called AFRTS, Armed Forces Radio and Television Service. The DEROS (Date of Estimated Return From Overseas) is when an Airman is expected to return to the US.

Educational opportunities are available on base, from local colleges offering classes to the Community College of the Air Force (CCAF). An accredited community college offering associate's degrees, CCAF was established in 1972. The training schools provided technical education, while general education came from regionally accredited civilian institutions. The Southern Association of Colleges and Schools accredited the college. Since 1976, CCAF has conferred over half a million associate degrees.

One unique educational opportunity is offered by the Air Force Institute of Technology (AFIT). From its main campus at Wright-Patterson AFB, Ohio, AFIT provides fully accredited, defense-focused graduate education and research in engineering, management and logistics. People may also be selected to attend a civilian university for graduate work, often in engineering or the sciences. Prospective USAFA or Professional Military Education (PME) instructors may attend an appropriate program to bolster their knowledge in a given subject. The Defense Language Institute offers foreign language training for members going to assignments where they will be deeply immersed in a foreign environment.

The Exceptional Family Member Program (EFMP) supports Airmen and their family members with special needs. These situations are taken into consideration during the assignment process. EFMP provides family support, referrals and resources, medical information, special educational and legal services, and assistance in obtaining these services in local communities.

HOW'S THE FOOD?

Air Force Dining Facilities (DFAC) are usually excellent. Visitors are often impressed by the quality and variety of DFAC meals. Food Service Craftsmen take pride in their work and are offended if you call it a chow hall. Airmen of

JET NOISE IS THE SOUND OF FREEDOM

a certain age, soldiers and Marines may use those terms out of habit.

Base Operations often had a short-order kitchen, but that has become rare. Flight Kitchens (run by the DFAC) provide box lunches for flight crews, maintainers and passengers. Those meals have the unflattering nickname Box Nasty. It's typically a lunch meat sandwich, or the trusty peanut butter and jelly sandwich, usually presented with a piece of fruit, a bag of chips, a can of soda, a half-pint of milk and a treat—maybe a candy bar and a packet of snack crackers. Quality varies.

Possibly the best snack bar was at Tinker AFB, Oklahoma, known for the Tinkerburger, with its sesame seed bun, lettuce, tomato, pickle and sweet onions. Delicious. If you're looking for a quick bite nowadays, the DFAC can serve you a burger, but national chains may be available on base. These vary from base to base. Bowling alleys on base usually offer fast food.

Walking around, you may see signs designating leadership positions. The Air Force denotes unit leadership and staff positions differently than the other services. Above wing level, you see the more widely used military staff system (A1 = personnel; A2 = intelligence; A3 = operations; A4 = logistics, etc.). The Air Force traditionally uses letters at wing level and below, although this is changing. CC is unit commander; CV is vice commander; CD is deputy commander; CCC is Command Chief; CCS is commander's secretary; CCE is executive officer; CCF is First Sergeant. Other areas may include MX (aircraft maintenance), OPS (Operations), LG (logistics), MS (mission support), and MED (medical). The Operations Group commander would be OG/CC, and the Maintenance Group commander is MXG/CC.

Command Post might be your first stop on a base. It is a 24/7/365 operation, where someone answers the phone even when the commander and most of the team are asleep or away. If there is an alert mission, the Command Post has responsibility for mission notifications.

The wing headquarters will include Judge Advocate (JA), individually and collectively called JAG (Judge Advocate General). They're the lawyers on base. SAPR (Sexual Assault Prevention and Response) and SARC (Sexual Assault Response Coordinator) are in the wing staff. The military takes sexual assault seriously. It is usually punished to a degree much greater than in civilian life. If a civilian man puts a hand on a woman's bottom without permission, he might get slapped, and possibly she files an assault charge. Do it in the military, and you'll be registering with the police as a sex offender for the rest of your life.

The wing staff also includes Intelligence, Chaplain, Equal Opportunity, Public Affairs and Inspector General (IG). Intelligence at the wing level collects information from personnel in combat and disseminates higher headquarters material pertinent to the unit's mission. Air Force chaplains are often identified by the letters Ch before their rank. They are religious ministry professionals who support the spiritual needs of Airmen. They lead religious services, provide personal guidance, and consult with leadership on moral and ethical issues. Equal Opportunity ensures fair treatment, and the IG investigates complaints. Public Affairs deals with civilian media inquiries and sends out press releases and other data to the community and the general public.

One group you may see at the Command Post is the CAT—Crisis Action Team. A CAT comprises personnel focused on mission execution. If it's a bad day, you might see how an incident affects the mission. Example: A mishap closes a taxiway. There will be an incident commander to deal with the immediate situation. However, the wing commander will have to decide how to launch and recover planes while fire trucks, ambulances, and aircraft maintenance work to resolve the mishap.

The Mission Support Group (MSG) has a broad mission. It includes Communications, Force Support, Aerial Port, Logistics Readiness, Civil Engineering, and Security Forces. Many of these functions are similar to those of a small town.

Force Support Squadron (FSS): Once called Services, this includes personnel (the human resources department), dining and lodging, mortuary and laundry, recreational and athletic facilities, family support activities, educational activities, child care and recreation facilities. Base Operating Support (BOS) includes services specialists, ground transportation and personnel specialists. The flying mission couldn't be sustained anywhere without these people on the ground. Prime RIBS (Prime Readiness in Base Services) provides lodging, food service, laundry, fitness and a mortuary that can be deployed quickly. The personnel accountability function is handled by PERSCO (Personnel Support for Contingency Operations). The PERSCO team provides the deployed unit with reception in-processing, force accountability, casualty reporting and limited personnel program support. So, Prime RIBS will feed you and provide a place to sleep, but it may be an MRE and a TEMPER Tent (Tent, Expandable, Modular, Personnel). The Meal Ready to Eat (MRE) is a self-contained meal. Some are tasty and some are not. Chili Mac, chicken burrito bowl, and beef stew are recommended. The TEMPER tent has been home to much of the Air Force at some point. Placed in various configurations, they create offices, medical facilities, workspaces and living quarters in deployed locations.

These functions can make a Bare Base operable. The naked truth about a Bare Base is clothed in post-Cold War history. For generations, Airmen looked at runways and long, straight stretches of highway and prepared to fight from there, assuming that much of what they needed to support the mission was available. However, after the Cold War, the Air Force became "expeditionary," with a requirement to deploy to a site with just a runway (even a dirt strip), a parking area capable of supporting aircraft and a source of water that can be made potable. Mobile facilities kits are stored as War Reserve Materiel, and civil engineers are trained in their use. This capability brings in air-transportable shelters, workspaces, kitchens and quarters, rendering any captured runway—or even a straight stretch of highway—an operable airfield.

The role of Civil Engineering (CE) cannot be overstated, whether in an expeditionary location or in garrison. Few Air Force organizations have a task list as diverse as that of a CE Squadron (CES): Infrastructure (Water & Fuels, HVAC, Pest Management), Facilities (Electrical & Power Production), Heavy Repair (Pavements & Structural); Engineering Project Execution and Portfolio Optimization; Fire & Emergency Service; Explosive Ordnance Disposal; Installation Management Flight (Environmental Compliance, Military Housing, Privatized Housing, Dorm Management); and Readiness & Emergency Management (Expeditionary Engineering, Emergency Management).

Some key functions of CE are Firefighting, Prime BEEF, RED HORSE, CBRNE, and EOD.

Crash Fire Rescue (CFR): Fire departments on Air Force bases are focused on the runway environment, but respond to structure fires, wildfires and situations where a civilian community might call for a paramedic or EMT. They are qualified on firefighting equipment and tactics, and trained in cardiopulmonary resuscitation and emergency medical technician skills.

PRIME BEEF is Prime Base Engineer Emergency Force, the team that provides highly skilled, rapid-reaction civil engineer forces to support contingency and installation sustainment missions. To open and operate a base, you need electricians, plumbers and all sorts of trained craftsmen.

RED HORSE is an acronym for Rapid Engineer Deployable Heavy Operational Repair Squadron Engineers. In 1965, the Air Force started its first heavy-repair civil-engineering squadrons, similar to Navy Seabees. These provide the USAF with a mobile, rapid-response, civil-engineer capability. These 400-person units can rapidly deploy to support contingency operations, even in high-threat environments. They deploy with their own weapons, equipment and even food service and medical support. There are even airborne qualified RED HORSE teams, primarily for airfield repair. They do runway and ramp construction, maintenance and repair, and can take on any civil engineering task, from damage assessment to building construction. RED HORSE members wear red hard hats or red baseball caps.

CBRNE (Chemical, Biological, Radiological, Nuclear and Explosives) specialists are the people who ensure that Airmen can survive and operate in any environment. Survivability involves hardening facilities, decontamination and mitigating the effects of various weapons, including blast, fire, radiation and electromagnetic pulse, by minimizing exposure and providing medical care for those affected.

EOD (Explosive Ordnance Disposal) personnel are the bomb squad. EOD personnel support civilian authorities with bomb problems, teach troops about unexploded ordnance (UXO) hazards, and aid law enforcement. EOD is known for saying their job is one of initial success or total failure.

Civil engineers build and maintain facilities and utilities, perform surveying and planning, and are the focal point for explosive ordnance disposal and disaster preparedness. The civil engineers who build and repair roads, buildings and infrastructure are often called "Dirt Boyz." They can create a base where none existed and put up tent cities to shelter and sustain thousands of Airmen. They build and repair runways. They provide light, heat, electricity and water. They are vital.

The Aerial Port Squadron (APS) is the domain of those who handle cargo processing, aircraft loading and unloading, load planning, inspection and load-team supervision. They provide an air terminal operations center for passengers and cargo. "Port Dawgs" do all this. Without aerial ports and the troops who run them, the airlift mission would not happen. Officially known as Air Transportation Specialists, they plan, schedule and process passengers, cargo and baggage and load or unload all of these. This unsung team makes the Air Force's global reach possible.

Logistics Readiness Squadron (LRS) combines supply, transportation, and fuels capabilities. They provide transportation of personnel, household goods, aircraft parts and cargo. The POL (Petroleum, Oil, Lubricants) section manages storage, transportation, distribution and quality of jet fuel, oil, gasoline and cryogenics, providing liquid oxygen and nitrogen. LRS manages plans and implements unit mobility taskings, such as emergency response deployments, combat support capability, and expeditionary support. Transportation troops manage the unit's vehicles and supply specialists manage and store mission-critical materiel. These are known as Government Owned Vehicles (GOV). Most Air Force people will drive a GOV at some point in their careers. A privately owned vehicle (POV) is yours, and it's important to note that there are areas on a base where signs say "No POV." So don't drive your personal vehicle there.

Security Forces Squadron (SFS) provides security for all personnel, resources, and other base assets assigned to the unit. Known as Defenders, Security Forces personnel proudly wear dark blue berets with the words "Defensor Fortis," meaning "defender of the force." Their work stretches from law enforcement to base defense. Central Security Control (CSC) houses SF communications equipment, sensor system displays, and maintains detailed maps of the base and its surrounding area. The Base Defense Operations Center, usually called BDOC (pronounced "bee-dock"), is staffed by at least two Defenders at all times. They monitor security systems for base facilities, communicate with security patrols and monitor flight line activity and traffic on base. One special group of Defenders is the RAVENS. The Phoenix Raven program was implemented in 1997, to put Security Forces aboard airlift aircraft for security if the flight is bound for unfriendly or unsecured destinations. Ravens work in small teams—two to six members—and provide security at the airfield as well as on-board security, similar to federal air marshals. You may hear them called "sky cops," but they don't like that. Defenders are the closest thing the Air Force has to an infantry. Defenders ensure the safety of the installation and its personnel. These Airmen were originally called Air Police (AP) from 1948 to 1966, then Security Police (SP) from 1966 to 1997. They are now called Security Forces (SF).

Communications Squadron provides communications services and customer support. The unit provides trained and ready communication-information professionals in support of air and space expeditionary force requirements. They provide telephone, internet, computer, and cyberspace support. There are two major computer networks. SIPRNET (Secret Internet Protocol Network) is for classified messaging and email, while the unclassified network is NIPRNET (Non-classified Internet Protocol Router Network). Comm may provide Information Operations (IO), allowing friendly forces to collect, control, exploit and defend information across the range of military operations, from peace to war. This includes defending one's own information while exploiting that of opposing forces. Comm also makes Command and Control (C2) possible. The military must move information, orders and resources to accomplish its missions. There is C3—command, control, and communications; C3I—command, control, communications, and intelligence; C4—command, control, communications and computers; and C4ISR—Command, Control, Communications, Computers, Intelligence, Surveillance and Reconnaissance. C2 is often used as shorthand for all of this. Comm's catchphrase, "No Comm, No Bomb," sums up the importance of their work.

Contracting Squadron delivers contract support to deliver timely acquisition capabilities and sound business decisions.

Comptroller Squadron handles the unit's financial management, manages all base appropriated funds and oversees non-appropriated funds for the wing. It develops and executes the unit's financial plans. More interesting to other Airmen, they take care of basic pay, bonuses, allowances, travel pay, hostile fire pay, flight pay, etc. Be nice to the people who keep your pay straight! One of the more unusual pays is Foreign Language Proficiency Pay (FLPP). The military needs people who speak foreign languages and will pay bonuses for this skill. Your two years of high school French aren't enough. You must be proficient in reading, writing and speaking languages useful to the US military, such as Russian, Chinese, Arabic or others.

There is a Medical Group at the base hospital. Most of the functions you might find in a small-town medical facility are present. The hospital exists to serve military needs but also cares for dependents and retirees. TRICARE is military-sponsored medical insurance, and major medical conditions may require off-base medical practitioners or referral to a major USAF facility such as Wilford Hall Medical Center in San Antonio or other regional military medical facilities. There is a dedicated Flight Medicine section, unlike most civilian hospitals. Things that might not matter to most people, such as bad sinuses, could be career-ending for a flier. Smart alecks call the hospital the "Medical Hobby Shop," which comes from the base Auto Hobby Shop, as if it were another place for amateur fix-it guys.

Moving toward the flightline, we see the Maintenance Group facilities. This is where aircraft are repaired and prepared for flight. Maintainers—aircraft maintenance troops—are meticulous, quietly dedicated and professional. Their jobs are literally matters of life and death. They take pride in never sending a pilot or aircrew out in a jet they wouldn't fly aboard themselves. Fliers get the glory but the Air Force doesn't turn a wheel without maintainers. One thing will always be true about their work: "Pilots without maintainers are just pedestrians with expensive sunglasses and cool leather jackets."

It's important to note that the strategic missile forces that provide so much of the nuclear deterrent have similarly dedicated maintainers. The missile operations and maintenance teams have a demanding job, balancing the awful burden of their mission with the hope that they never have to carry it out.

There is a Maintenance Operations Flight (MOF) responsible for running the Maintenance Operations Center (MOC—a maintenance command post), Plans, Scheduling and Documentation (PS&D), Engine Management (EM), Training Management, Analysis, Budget Analysis and Plans and Programs.

The Maintenance Squadron (MXS) consists of specialists, organized into flights: Propulsion, Avionics, Accessories, Aerospace Ground Equipment (AGE), Pneumatics, Hydraulics, Avionics, Guidance and Control, Fabrication, and other flights (such as munitions or electronics) as required.

The Aircraft Maintenance Squadron (AMXS) is responsible for aircraft generation, including servicing and inspection, as well as maintaining, launching, and recovering assigned aircraft.

Depending on the unit's mission, there may be Component Repair, Electronics, or Munitions Maintenance Squadrons.

Quality Assurance (QA) is the Maintenance Group commander's technical adviser and evaluates the quality of maintenance accomplished, while managing the Maintenance Standardization Evaluation Program. Aircraft Maintenance is one of the most disciplined and exacting fields in the Air Force. Everything maintainers do is governed by Air Force Technical Orders (AFTO). Technical Orders, often called "TOs," provide instructions for the safe and effective operation and maintenance of aircraft, systems and other equipment. Time Compliance Technical Orders (TCTO) provide immediate or urgent instructions for aircraft or equipment modifications, repairs or inspections.

One of the most frustrating things maintainers encounter is the dreaded "Could Not Duplicate" (CND). Sometimes, when a flight crew writes up a problem, maintenance can't replicate it; therefore, it can't be fixed. CND is the term used in the forms where corrective action would be noted.

Going on the flightline is governed by red lines on the ground. You enter where you're told to. The restricted flightline area is bordered by a red line on the ground around the aircraft parking area. You enter and leave at a designated point. Crossing the red line, known as "Breaking Red," is a major faux pas and a security violation. You will find yourself face down on the tarmac, "kissing the pavement," until someone excuses you. This is much more fun for the guys with the guns than for the poor soul on the ground.

AIR FORCE PILOT TRAINING lasts about a year, and is currently conducted at three locations: Vance AFB, Oklahoma; Columbus AFB, Mississippi; and Sheppard AFB, Texas. Trainer aircraft have long lives. They are acquired in large numbers, built for reliability and safety and replacing them is often a low priority. The **T-33** was a variant of the P-80 fighter plane, which was being phased out of frontline service by the end of the Korean War. However, the T-33 proved to be an excellent trainer and was used for pilot training into the 1960s, remaining in service at the Air Force Academy until 1975. A few stayed around the Air Force till 1997. The "T-Bird" was replaced by the **T-37 TWEET** and the **T-38 TALON** for basic and advanced pilot training. T-37s trained pilots from 1956 until 2009. The little (how little? One step up, and you were in the cockpit because the canopy rail was only about four feet off the ground) Cessna aircraft was easy to fly, maneuverable and aerobatic. It didn't have a great deal of power and might be best remembered for the deafening shriek of its engines. Its nickname, the "6,000-pound dog whistle," was fitting given its ability to turn jet fuel into noise. Pilots moved into the T-38 Talon after the Tweet. The Talon has been used to train pilots since 1961. Its stubby wings and high speed mark it as a product of its time, when the phrase "speed is life" was Air Force gospel. Pilots marveled that you could roll it through 720 degrees in one second. The last T-38 was delivered in 1972, and it's still training fighter-track pilots. Test pilots and flight test engineers are trained in T-38s at the US Air Force Test Pilot School at Edwards AFB, and AFMC (Air Force Materiel Command) uses the T-38 to test experimental electrical and weapon systems. NASA uses T-38s as trainers and as chase planes.

In January 1992, the Air Force accepted the first production model **T-1A JAYHAWK**, and the following year saw the first student sortie in the new trainer. In May 2000, the first production model **T-6A TEXAN II** aircraft arrived at Randolph AFB, Texas, replacing the T-37. Since the 1990s, pilot candidates started in the T-6 Texan II, then, if bound for airlifters and tankers, moved on to the T-1

Jayhawk, or to the T-38 if slated for fighters and bombers. The **T-7A RED HAWK** is expected to replace the T-1 and the T-38.

COMBAT SYSTEMS OFFICERS (CSO)—pronounced "sizz-o"—are the flight crew members who operate weapons, guidance and electronic warfare systems. CSOs can be bombardiers (formerly B-52 Radar Navigators, B-1 Offensive Systems Officers), Weapons Systems Officers in fighters (once called WSOs), Electronic Warfare Officers (erstwhile EWOs, Ravens, Crows, Defensive Systems Officers), and a very few are still navigators (B-52, RC-135, AC/MC-130) responsible for positioning the aircraft. Combat Systems Officers (CSO) are trained at Pensacola Naval Air Station, Florida.

Air Battle Managers (ABM) are trained at Tyndall AFB, Florida. They manage airborne radar sensors, electronic countermeasures and communications networks to direct and coordinate air operations. These are Rated Officers, those qualified in an aeronautical specialty (Pilot, CSO, ABM, RPA (Remotely Piloted Aircraft) pilots and flight surgeons). These officers can receive flight pay and other incentives for continued service. Most Air Force officers are "Non-Rated," without an aeronautical rating.

Enlisted fliers may train in various locations, depending on their assignment. They can become Career Enlisted Aviators (CEA), Airmen whose career field is Loadmaster, Boom Operator, Flight Engineer, Gunner or various crew duties in Reconnaissance and Electronic Combat aircraft.

Officers come from the Air Force Academy, Air Force ROTC or Officer Training School. Some are prior enlisted, but this is not required.

THE AIR FORCE ACADEMY

In 1954, President Eisenhower signed legislation creating the United States Air Force Academy (USAFA) at Colorado Springs, Colorado. Lt. Gen. Hubert R. Harmon was the first superintendent. The Cadet Honor Code was created by the Class of 1959 and remains a simple statement: "We will not lie, steal, or cheat, nor tolerate among us anyone who does." Hollywood director Cecil B. DeMille designed the cadet parade uniform. The first cadets, the class of 1959, arrived in 1955. Since then, the Air Force Academy has commissioned more than 50,000 officers. Nearly 200 Academy graduates have been killed in action, and hundreds more have died in the line of duty.

The Air Force Academy is a challenging school with a rigorous academic program. About 10,000 people apply each year, and about 1,400 are accepted. Graduating classes are typically around 1,000 cadets. The USAFA Preparatory School (the "Prep School") is an intense academic program (particularly in math and science) to prepare civilians and enlisted members who did not receive appointments for the Academy.

Academy cadets are sworn into the military and receive six weeks of Basic

Cadet Training before starting academic work. Engineering, calculus, physics and chemistry courses are required. About half the cadets major in technical subjects. Graduates receive a Bachelor of Science degree and a commission as second lieutenants in the US Air Force or Space Force. Approximately 900 USAFA graduates have become generals, a rate significantly higher than that of officers from other commissioning sources. Some 40 astronauts, six Air Force Chiefs of Staff and nine Vice Chiefs of Staff are USAFA graduates.

Cadets pay no tuition, but receive room and board, as well as a monthly stipend. Those who don't graduate are expected to serve a term of enlistment or repay their tuition. Upon graduation, they will serve at least five years of active duty as Air Force or Space Force officers.

The Academy is sometimes called the Zoo, or The Blue Zoo. Cadets are jokingly called Zoomies. First-year cadets are called Doolies by everyone.

An Academy candidate must be nominated by a Member of Congress or a senior government official. Children of military personnel who died in the line of duty or were 100 percent disabled, and those of Medal of Honor recipients, may receive special nominations.

Air Force Reserve Officers Training Corps (AFROTC) is the largest source of Air Force, Space Force and Air Reserve Components commissioned officers. Gen. Dwight Eisenhower established Army Air Forces ROTC units at several universities after World War II. Today there are more than 140 AFROTC detachments. Cadets enroll in military training in addition to attending their college courses. They take military classes for four years, attend training after the sophomore and junior years and are commissioned on graduation.

Officer Training School is available for individuals who have a college degree. It is often a path to commissioning for enlisted Airmen, but it accepts people with no military experience for a nine-week program leading to a commission.

Promotions are a sensitive topic in the military. It is frowned upon to say, "I want to be promoted," but everyone wants to be promoted. For an officer to be promoted early to O-4, O-5 or O-6 is called Below Primary Zone (BPZ). Getting ahead of one's peers indicates that the officer is on track for high rank. On the enlisted side, the Below The Zone (BTZ) promotion from E-3 to E-4 receives that stripe six months before the regular time-in-service promotion.

Another career-enhancing opportunity is the By Name Request (BNR). A senior officer or civilian leader may create a BNR for certain positions, such as Aides, Executive Officers and other staff positions, to place trusted people in his retinue. A BNR is a career boost.

One aspect of an Air Force career is "The Board." An Airman will face many different boards. Many are positive: Promotion, Selection (for schools, command or training opportunities), Awards (Airman of the Quarter/Year,

Command Selection, etc.). Some boards are negative: Flying Evaluation (an FEB can end your flying career); Selective Retention (can end your Air Force career); Force Shaping (downsizing); Medical, etc.

The Enlisted/Officer Performance Report (EPR/OPR) is the official record of an Airman's career. These document performance and provide information for major career opportunities, such as promotions, assignments and schools. The rules and rubrics for these appraisals have undergone numerous changes over the years. The brief statements describing an Airman's work are called "bullets." Officer promotions are based on the Promotion Recommendation Form (PRF). It is a one-page story of a career. The rules change often—civilian education can or can't be mentioned; the way one is compared to peers (stratification) may change. The constant is that everyone wants a "Definitely Promote" recommendation, and these are limited. A "Do Not Promote" is a kiss of death. Most people receive a "Promote." This means it is highly competitive to get the best possible recommendation. Most officers will make major, about half make lieutenant colonel, and half of the lieutenant colonels will make colonel. Promotion to brigadier general will only occur for about 2 percent of colonels. Passover is a term that is dreaded among Air Force officers, even though it is likely for almost all at some point. Being passed over for promotion is a terrible experience. For officers, being promoted after a passover is unlikely. Passed-over officers are seldom given career-enhancing opportunities; these are typically reserved for individuals who may be considered for future promotions. Some officers come to grips with this disappointment; for others, it is a lifelong heartache.

The Weighted Airman Promotion System (WAPS) governs enlisted promotions. It is scored by time in service, time in grade, decorations, EPRs and examinations on professional knowledge and job skills. The Airman receives a score, and the "cutoff"—the lowest score for promotion—is published, so individuals know how close they came to promotion. An enlisted member can compete repeatedly for promotion to the next rank. It is common to retire as a master sergeant, although some will finish as Technical or even Staff Sergeants. This is sometimes a case of being trapped in a career field with limited opportunities. Rising to Senior or Chief Master Sergeant is a tremendous accolade, carrying great respect. Those ranks are limited to 2 percent and 1 percent, respectively, of the enlisted force.

FAMOUS ACES

Fighter Aces are the best known heroes in USAF history. In the World Wars, thousands of fighter pilots downed five or more enemy aircraft in battle, earning the title "ace." Capt. Eddie Rickenbacker was America's top ace with 26 victories in the First World War, and Maj. Dick Bong shot down 40 Japanese planes in World War II. In air-to-air combat in Korea, Capt. Joe

McConnell was the top-scoring American, with 16 victories. There were a few fighter aces in Vietnam, all of whom flew the F-4 Phantom. On August 28, 1972, Capt. Steve Ritchie became the first USAF ace of the war. In September, WSO Capt. Charles B. DeBellevue got two kills in one day to become an ace, and with victory number six, he became the highest-scoring American ace of the war. WSO Capt. Jeff Feinstein got his fifth MiG on October 13, 1972.

Since Vietnam, there have been no aces, but a few USAF pilots have shot down three enemy aircraft in combat: Col. Cesar "Rico" Rodriguez shot down two Iraqi MiGs in Desert Storm, then got a Serbian MiG-29 in the Kosovo campaign. Capt. Thomas Dietz destroyed two MiG-21s during Desert Storm, then downed an Su-22 during Operation Provide Comfort. First Lt. Bob Hehemann matched that total with two Su-25s and a PC-9 downed in those operations. Capt. Robert "Wilbur" Wright shot down three Serbian Galeb fighters over Bosnia in 1994, the first three-kill day by an American since Joe McConnell did it in 1953.

To be an ace is a rare distinction, even among fighter pilots. Only about 1 percent of all fighter pilots achieve this status. A handful of Airmen have been aces in two wars. Several World War II aces achieved this in Korea, too. They were:

George Davis Jr. was the first USAF two-war ace, destroying seven Japanese aircraft during World War II and fourteen enemy airplanes in Korea. Davis flew P-47s on 266 missions from New Guinea and the Philippines. On February 10, 1952, Davis got his last two MiGs as he and his wingman stormed into a dozen of the enemy to break up their ambush of some F-84s. The bold attack scattered the enemy, but Davis's plane was severely damaged and plunged to the ground. George Davis was killed, but his body was never recovered. He was awarded the Medal of Honor.

Francis Gabreski—inevitably nicknamed "Gabby"—was the son of Polish immigrants. He won his wings in 1941 and was assigned to Wheeler Field, Hawaii, where he survived the Japanese attack on December 7, 1941. He went to England and flew with the Polish squadrons of the RAF to learn from their experience. Gabby joined the 56th Fighter Group in 1943 and became the most successful American fighter pilot in the European Theater, shooting down 28 German aircraft. On July 20, 1944, with his bags packed to go home, he couldn't resist the chance to fly just one more mission. Strafing a German airfield, his P-47's propeller contacted the ground, and he crash-landed. Taken prisoner, he encountered Hanns Scharff, Germany's most successful POW interrogator. Scharff greeted him with a smile: "Hello, Gabby! We've been waiting for you!" Gabby spent the rest of the war in a POW camp. He fought in Korea and became a jet ace with 6.5 victories, the first USAF two-war ace. With 34.5 victories, he was America's top living ace till his death in 2002.

Vermont Garrison was a schoolteacher in Kentucky before World War II—an unlikely hero. He washed out of AAF pilot training but joined the RAF and flew for the British until he transferred into the AAF. He scored 7.33 victories before his P-51 was shot down while defending bombers. He spent over a year in a German Stalag. Continuing his Air Force career, he was assigned to Korea and shot down 10 MiG-15s. He flew his last combat missions in the F-4C in the Vietnam War.

James Hagerstrom flew P-40 Warhawks during the New Guinea campaign. He destroyed six Japanese aircraft, including four in one mission. He returned home in February 1944 and married a WASP (Women's Airforce Service Pilot). Hagerstrom flew with the Air Guard and was recalled to active duty for Korea in 1952. In six months in MiG Alley, he shot down 8.5 enemy airplanes, the last on the day he was scheduled to depart Korea. Hearing that a pilot was needed, he hustled to the squadron and went out as part of a four-ship formation. He got his last victory, dressed not in a flight suit but in the dress blues required for travel as a passenger on a military aircraft.

Harrison Thyng graduated from the University of New Hampshire and earned his wings in 1940. He went overseas in 1942, flying Spitfires, and was credited with five kills before he was wounded in North Africa. He recuperated and became the commander of a P-47 fighter group in the Pacific. In 1951, he deployed to Korea and downed five MiGs, bringing his two-war total to 10 victories. Thyng engaged in aerial combat against pilots from seven countries: Germany, Italy, Vichy France, Japan, China, the USSR and North Korea.

William Whisner downed 16.4 (sometimes pilots shared credit) German aircraft, six of them over Merseburg, Germany, on November 21, 1944. On January 1, 1945, he met an attacking force of German fighters. The fight began just after takeoff from a field in Belgium, only 1,500 feet above the ground. Whisner got four of the enemy, but damage forced him to land a few minutes later. Whisner, 21 at the time, was awarded the Distinguished Service Cross for each of those missions. In 1951, Whisner logged six victories in Korea, earning a third DSC for saving another pilot under attack. Whisner died in 1989 from complications following a yellow jacket sting.

JET NOISE IS THE SOUND OF FREEDOM

THINGS WE SAY: "COMMON KNOWLEDGE"

"What's the difference between God and a fighter pilot? God doesn't think he's a fighter pilot."

How do you know if there is a fighter pilot at your party? He'll tell you.

Difference between a fairy tale and a war story: Fairy tales begin, "Once upon a time," but war stories start out with, "This is no sh!t!"

Peace on earth—through superior firepower.

Our sense of humor is different. So is our language. For example, the Army uses the word "Post" to describe its installations, while the Air Force says "Base." Many of the things we say are common in the entire military, and some have become common usage. But you can usually tell if someone has served by these things we say.

ABOVE MY PAY GRADE: A common response when someone complains about a decision made by more senior people. It's a little more polite than "Whaddya want me to do about it?!"

AIRCRAFT IDENTIFICATION CODES: Aircraft are identified with a simple code: A—Attack; B—Bomber; C—Cargo; F—Fighter; H—Helicopter; K—Tanker;[21] L—Liaison; O—Observation; P—Patrol; T—Trainer; U—Utility; X—Experimental; Y—Prototype.

It gets complicated when we try to explain the aircraft's mission further. For example, the C-130 and C-135 cargo planes have served in many other roles: AC (attack gunship); DC (drone control); EC (electronic); HC (personnel recovery support); JC (satellite film cartridge retrieval); KC (air refueling); LC (ski-equipped 130s); MC—Multi-Mission Cargo; NC (flight test); RC (reconnaissance); TC—Trainer; VC (VIP); WC/WB (weather reconnaissance). There is AT (attack/trainer); CV (cargo/vertical/short field); EB/EF (electronic bomber/fighter); FB (fighter/bomber); OA (observation/

---

[21] Most of these are the first letter of the mission. But K? Some say it comes from kerosene, a primary ingredient in jet fuel. But the use of the K for the tanker predates jet aerial refueling, as in the KB-29. K was likely just an available letter, since T (for Trainer) was already in use.

attack); OV (observation/vertical/short field); RB (reconnaissance/bomber); SR (strategic reconnaissance); MQ/RQ (Remotely Piloted Aircraft).

In the early days of aviation, manufacturer names were used as identifiers: DH (DeHavilland); SPAD (Société Pour L'Aviation et ses Dérivés); JN (from the Curtiss J and N trainers, often called "Jenny"); LUSAC (LePère United States Army Combat); and MB (Martin Bomber). During World War II, BT, PT and AT were Basic, Primary and Advanced Trainers, while P stood for Pursuit and F stood for photographic reconnaissance.

AIRCRAFT NICKNAMES

Most Air Force aircraft have formal names: the F-16 is the Fighting Falcon, and the C-17 is the Globemaster III. But Airmen refer to them as the Viper and the Moose. Many planes get nicknames from the people who maintain and fly them:

A-1 Skyraider: Sandy, Spad.

A-7 Corsair II: SLUF (Short Little Ugly Fella).

A-10 Thunderbolt: Warthog, Hog.

AC-47: Spooky, Puff the Magic Dragon.

AC-119G: Stinger.

AC-119K: Shadow.

AC-130A/E/H: Spectre.

AC-130U: Spooky.

AC-130W: Stinger.

AC-130J: Ghostrider.

B-1 Lancer: Bone (B-One). Makes sense when you see it in writing, doesn't it? Officially, the Air Force calls the B-1 "Lancer." I've never heard a crewmember use that.

B-2 Spirit: Bat, Boomerang, two names that allude to the aircraft's shape.

B-26 Marauder: Widowmaker.

B-36 Peacemaker: "Six turning, four burning." It had six propellers and four jet engines. Also, Aluminum Overcast. It was a very large airplane.

B-52 Stratofortress: Buff (Big Ugly Fat Fella).

C-5 Galaxy: Fat Albert; FRED (Freaking Ridiculous Economic Disaster). Its maintenance costs are high, and it has a history of breakdowns. It's said that if you see three C-5s and two are on jacks, the base has only two sets of jacks. But when FRED flies, it's an incredible sight and a vital aircraft.

C-17 Globemaster III: Moose. Globemaster I was the C-74, and Globemaster II was the C-124.

C-47 Skytrain: Dakota, Gooney Bird.

C-119 Flying Boxcar: Dollar Nineteen.

C-121 Constellation: Connie.

C-124 Globemaster II: Old Shaky.

C-130 Hercules: Herk.

C-141 Starlifter: Star Lizard, for its "lizard" camouflage.

E-3 Sentry: AWACS ("A-wax") From its Airborne Warning and Control System.

E-4 Nightwatch: Kneecap. Comes from an attempt to pronounce its role: NEACP (National Emergency Airborne Command Post).

EF-111 Raven: Sparkvark.

F-111: Aardvark. It never had an official name. Crews called it the Aardvark.

F-4 Phantom II: Rhino.

F-4G Wild Weasel: Weasel.

F-15 Eagle: The F-15 is a rare bird. Its official name is Eagle and that's what everyone calls it.

F-15E Strike Eagle: Mud Hen.

F-16 Fighting Falcon: Viper; Electric Jet; Lawn Dart.

F-22 Raptor. Like the Eagle, it wears its given name proudly.

F-35 Lightning II: Panther.

F-84 Thunderjet: Hog.

F-84F Thunderstreak: Super Hog.

F-100 Super Sabre: The Hun (short for 100); Lieutenant Killer. It had a reputation for being unforgiving and dangerous for new fliers.

F-101 Voodoo: One-O-Wonder.

F-102 Delta Dagger: Deuce.

F-104 Starfighter: Missile with a man in it.

F-105 Thunderchief: Thud.

F-106 Delta Dart: Six.

F-117 Nighthawk: Stealth.

H-21 Shawnee: Flying Banana. One look, and it's obvious!

KC-10 Extender: Gucci; Big Sexy.

KC-135 Stratotanker: Since about 1960, the KC-135 has been the primary tanker in service, so it's simply become the Tanker or Flying Gas Station. The KC-135A's water injection system augmented engine thrust and gave it the monikers "Steam Jet" and "Waterwagon."

LC-130 Hercules: Skibird

"Lead Sled." Many heavy, fast aircraft picked up this nickname, including the F-4 Phantom, F-84 Thunderjet, F-105 Thunderchief and SR-71 Blackbird.

P-26: Peashooter. The 1930s fighter had a long tubular gunsight that looked like a toy peashooter.

P-38 Lightning: Fork-Tailed Devil.

P-47 Thunderbolt: Jug. If it were standing on its nose, this World War II fighter resembled a milk jug.

RC-135: Hog Nose. Several models of RC-135 perform various reconnaissance missions. They look different, with many antennae, unusual paint schemes and fairings along the fuselage and tail. But they all have an elongated radome, giving them the hog-nose look.

SR-71 Blackbird: Habu.

T-33 Shooting Star: T-Bird.

T-37 Tweet: 6,000-pound dog whistle.

X-plane: An experimental aircraft: X-15, XB-70, X-1.

AIR FORCE GLOVES: Standing around with your hands in your pockets is a long-held military taboo. Other services mocked the Air Force for being lax on this rule. Finally, in 2021, the Air Force decided to permit it.

AIRMAN SNUFFY: An example of what NOT to do. Ironically, this is related to SSgt. Maynard "Snuffy" Smith, a World War II Medal of Honor recipient. Snuffy Smith was a B-17 gunner who fought off repeated fighter plane attacks, administered first aid to wounded crewmen and put out fires caused by enemy fire. His final act was urinating on a fire to extinguish it. However, Smith was hard to get along with and was demoted to private due to his behavior. After the war, he was in and out of petty trouble. Even brave heroes are flawed; sometimes, the most unlikely person saves the day. So, don't judge a book by its cover; take each person as an individual. Some of the best people I've ever known were Airmen, but not every Airman is a shining example. Don't be Airman Snuffy!

ALCON: It means "all concerned," and everybody receiving such a message is expected to heed its words. ALCON could be all the members of a squadron, wing or the whole Air Force.

ALL THRUST, NO VECTOR: Someone who is ambitious, helpful or busy but has no idea what they're doing. Going fast, but no idea where!

ARMY PROOF: The Army hates this one. You can make something foolproof, but can you make it Army Proof? To Army Proof it, simplify the instructions and add a picture. For example, the MRE heater unit comes with a sketch and instructions that show how to use it, including leaning the hot package against "a rock or something." It says that. The Claymore mine has "Front towards enemy" embossed on the front. Was that a designed feature or discovered after an unfortunate experience?

ATE UP: A situational term. You might say someone is "ate up" and overeager or excessive, but it can also mean they have a strong sense of responsibility, duty and service.

BAG DRAG: Loading/unloading your gear on an aircraft because you're going somewhere good is fun. But a Bag Drag to a different jet, because the first one you loaded is broken, is an event cursed by everyone.

BALLS TO THE WALL: To go as fast as possible. It refers to pushing the throttle levers of an aircraft up to the firewall, to maximum speed. The levers have a rounded ball on the top of them. Hence, the balls are at the wall.

BANDITS: Aircraft identified as an enemy.

BARBER POLE: The striped needle on some airspeed indicators that marks the maximum allowable speed. Being "on the barber pole" means going as fast as possible, short of overspeeding the aircraft, which can cause engine or structural damage and loss of control.

BASE X: The location of a story that the speaker doesn't mention—for many reasons—avoiding self-incrimination, protecting the guilty or not embarrassing someone. "Back at Base X, this one guy who was a goof.☒"

BATTLEFIELD AIR INTERDICTION: Use of air power to target enemy logistics, supply lines, and troop movements, before they can impact the battle.

BLUE FALCON: Someone who is untrustworthy and will do anything to get ahead. It is the worst insult in the Air Force. *Don't ever be a Blue Falcon.*

BLUE STEELY: A six-passenger blue Air Force pickup truck.

BOHICA: It means "Bend over, here it comes again." It's a response to unfavorable news, like it was a painful medical injection in the posterior.

BOUGHT THE FARM: A term used when a pilot is killed in an aviation accident. It originated when the government reimbursed farmers for crops destroyed by aviation accidents in their fields. Farmers would inflate the value of lost crops to the point that, in effect, the pilot "bought the farm."

CANX: Cancelled. It may also be abbreviated CNX. "That mission was CANX'd this morning."

CAS (Close Air Support): Airpower directly supporting ground troops in combat.

CHECK SIX: It means "look behind you" or "keep your eyes open." Six o'clock is behind you, and twelve o'clock is in front. This comes from the aviator's habit of constantly checking for threats from the rear. It's a useful phrase on the ground too, referring to watching out for those who might do you wrong.

CHICKS: Planes in formation with a tanker: "Tanker 51 proceeding with chicks in tow." (That is, in an air refueling formation)

CHRISTMAS TREE: A feature of SAC bases was that alert aircraft were parked with noses pointed toward the runway. From the air, you can see that it's shaped like a stylized tree.

CLEAN: For air traffic, it means "No radar contacts"; To aircraft in formation, it means "No visible damage"; To maintenance from an inbound aircraft, it means "No external stores." On climb out, it means a clean configuration (gear up, flaps up). It can also mean removing dirt and muck, as it does everywhere else.

CLEARED: Requested action is authorized.

CLEARED HOT: In an airplane, ordnance release is authorized. On the ground, "you have permission."

CLOSE OF BUSINESS (COB): End of the duty day. "I want this done by COB."

CLOSED FOR TRAINING: Many support offices close during the duty day "for training." This makes little sense to the Airmen whose job is training every day for their mission. It's infuriating to pull into a parking lot, needing to get something done, only to find there's not a car in the lot. No lights are on inside, only a sign that says "closed for training." Curse words follow.

COME TO JESUS: It isn't a religious admonition, but it is advice that it's time to straighten up and fly right. "After those two guys got arrested, the colonel had a 'come to Jesus' meeting with all of us."

CONUS: Continental United States. The CONUS is the 48 contiguous states and the District of Columbia. OCONUS is outside CONUS.

DEADSTICK: In aviation, it is an approach and landing without power. It occasionally is used to say you're exhausted: "I deadsticked it home last night."

DEFENSIVE COUNTER AIR: Operations conducted to protect friendly forces and critical assets from enemy air and missile threats.

DG: Distinguished Graduate. Typically, the top 10 percent of an Air Force school or training course are designated as such.

DIRTBAG AIRMAN (DBA): An Airman who does not represent the Air Force Core Values. The United States Air Force is elite. The work is challenging, the missions are demanding and important. But in any large organization, some are not as sharp as others. The Dirtbag Airman is sloppy, breaks the rules and doesn't care about his work. He looks like a bag of donuts in uniform, forgets to salute, ducks inside when the National Anthem plays, shows up late, leaves early and takes breaks when the workload is busiest. Airman Snuffy comes through sometimes. The DBA doesn't. *Don't be a DBA!*

DIFFERENT COLOR MONEY: Government money is appropriated for specific things. You can't use Military Personnel (MilPers) funds for Operations and Maintenance (O&M) purposes, such as airplane parts, etc. "Different color money" is a way of saying this.

DISGRUNTLED AIRMAN: It can be the anonymous complainer, the guy who is unhappy with something and elevates it unreasonably: "We didn't have chocolate ice cream in the DFAC, and a disgruntled Airman wrote his congressman." It happens. When things go wrong, this character is often blamed: "Who broke this?" "There were no witnesses. Appears to be a disgruntled Airman."

DOLLAR RIDE: A first flight, usually in a new aircraft type. In the 1920s, barnstormers sometimes took people up on quick sightseeing flights for a dollar.

DOT: It refers to how a distant aircraft looks on the horizon. "I'm a dot" means "I'm out of here."

DOUBLE DIP: Collecting two paychecks simultaneously. This applies to retired military members working in civil service positions while drawing a military retirement check. It also refers to Reserve personnel who receive a military leave benefit from their civilian employment, allowing them to receive income from two sources (military pay and civilian salary).

DRIVER: Pilot. "Eagle driver" is F-15 pilot; "Herk driver" is C-130 pilot. A "heavy driver" is the pilot of a big airplane—bomber, tanker, reconnaissance, or airlifter.

ELEPHANT WALK: is when a large number of aircraft taxi in sequence.

EMBRACE THE SUCK: Sometimes you have to endure undesirable situations or live with decisions that negatively impact you. In other words, don't worry about what you can't control.

ENGAGED: Maneuvering with the intent to kill, having made visual and/or radar acquisition of a target.

ENVELOPE: The maximum performance parameters of an aircraft. Flying at the edge of the envelope means operating the aircraft at its maximum speed, altitude, or performance capability.

ETS: Expiration Time of Service, meaning the end of one's commitment or enlistment: "You gonna reenlist?" "No, I'm out on my ETS."

FANGS OUT: It means to be aggressive. "I saw the MiG, and my fangs came out."

FAST BURNER: Someone who is rapidly advancing in rank.

FAT PILLS: Donuts.

FEAR OF FLYING: This psychological condition affects many people. It reaches the level of "manifestation of apprehension"—that is, a state of psychological anxiety, apprehension and/or physical impairment. Many jobs in the Air Force involve some physical risk. Not everyone is cut out for every job. Fear of flying is disqualifying for those whose duties require frequent and regular flight. The British first diagnosed it during World War I. Perhaps the most famous Airman to battle this was Hap Arnold, who removed himself from flying for a few years after a couple of close calls.

FEAR, SARCASM AND RIDICULE: Someday, historians may prove this was a teaching methodology throughout the Air Force. For now, it's just an opinion.

FEET WET/DRY: describes a flight position over water or over land.

FIGMO: "Forget it, I got my orders." When a person doesn't care what's happening at this unit in the future because he's moving on, he's said to have gone FIGMO. A bad attitude.

FIGHTER PILOT BREAKFAST: At one time, it was a Coke and a smoke; Chuck Yeager said it was a cup of coffee, two aspirin and a puke. Now, it's an energy drink or a protein shake.

FIREWALL IT: "Speed up!" Piston airplanes had a firewall between the engine and the pilot, so pushing the throttles all the way forward to the firewall meant maximum acceleration.

FLEXIBILITY IS THE KEY TO AIRPOWER, AND INDECISION IS THE KEY TO FLEXIBILITY: The Italian theorist, Giulio Douhet, said that "flexibility is the key to airpower." Many frustrated and tired Airmen have added this coda while awaiting instructions following a mission change or some unexplained delay.

FLIGHT DOC: Flight Surgeon.

FLIGHTLINE: Where the airplanes are parked.

FOUR FANS OF FINANCIAL FREEDOM: A Reservist's term for flying big jets and picking up Reserve pay.

FORM 1: Toilet paper. The Air Force has an endless number of "Forms." Some wit decided that Form 1 must have been toilet paper.

FORMER MAJOR COMMANDS: MAC (Military Airlift Command); SAC (Strategic Air Command); TAC (Tactical Air Command). Even though these have been gone for over thirty years, they are remembered with some reverence, even by people too young to recall them.

FOX: "Missile away."

FPCON (Force Protection Conditions): These defensive postures may change due to local threats or significant events. Within each FPCON, various measures can be used to enhance force protection. These FPCONs go from NORMAL through ALPHA, BRAVO and CHARLIE to DELTA (an attack has occurred or intelligence indicates that terrorist action is imminent).

FRAUD, WASTE, AND ABUSE (FWA): This is misconduct involving government resources. Fraud is intentionally cheating the government to gain a financial benefit. Waste is the mismanagement of resources. Abuse is following questionable practices or policies that lead to unnecessary costs or expenses.

FURBALL: An air battle involving multiple aircraft with hostiles and friendlies mixed.

GIVE IT BACK TO THE TAXPAYERS: You can have it, I don't want it. "When the fire got to the cockpit, he decided to give it back to the taxpayers."

GOAT ROPE: A badly messed-up situation. I've never had to rope a goat, but it must be a wild scene: "Today's mission was a real goat rope!"

GO BALLISTIC: To lose your cool: "The colonel went ballistic when I showed up late for the briefing."

GO UGLY EARLY: It sounds like crude advice from drunken pilots at a bar, but it's the use of A-10s to support ground troops early on in a battle. Don't wait till you're in trouble to get close air support.

GONE WEST: It means a flier has died: "He survived the war, but he's gone west now." It may also be said, "flown west."

GOOD PEOPLE: A high compliment, even of one individual: "He's good people."

GOT YOUR SIX: It refers to fighter pilots constantly clearing behind them, ensuring no threat is approaching from six o'clock, the rear. It's the Airman's equivalent of "I've got your back."

GOVERNMENT TRAVEL/PURCHASE CARD (GTC/GPC): The GTC is for individual use. Many Airmen travel frequently, and people in these roles are issued a credit card for this official travel. The opportunity for misuse exists and it occasionally makes the news. The GPC is a commercial charge card used by units for small purchases, like an electronic petty cash. Unfortunately, this program occasionally makes headlines when some DBA (Dirtbag Airman) uses a GPC for shady things—bars, booze, nightlife, etc. Misusing the GPC or GTC is a serious thing. Don't do dumb stuff!

GREATEST TEAM IN THE WORLD: World War II AAF recruiting posters encouraged young men to join "The Greatest Team In The World." The Air Force is still that.

GROUND POUNDER: Non-flier.

GRUNT: Army types. All of 'em.

HAIR ON FIRE: It means to act with recklessness or a lack of control. "I rolled in with my hair on fire and overshot the target."

HANGAR #: A nonexistent hangar (e.g., if there are four hangars, then it would be Hangar 5), sometimes used over radios as code for the bathroom or a portable toilet. Example: "I need to run over to Hangar 5!"

HEATSEEKER: A Sidewinder missile, which homes in on heat sources.

HIGH DRAG: This has two meanings. One is a bomb dropped from low altitude with devices to slow it down. The other is someone who slows down the team, not the sharpest knife in the drawer: "He's a nice guy, but he is high drag."

HOLDING HANDS: It means aircraft are in visual formation.

HURRY UP AND WAIT (HUAW): The classic story of military life. My experience was that it's usually "wait and then hurry up."

I'D RATHER BE LUCKY THAN GOOD: It means that you realize that good people, even the experts, have bad days. So, being lucky is a good thing, especially when you've just walked away from something that could've gone very wrong.

IF I TOLD YOU, I'D HAVE TO KILL YOU: A humorous way of fending off questions about sensitive or classified information.

IF THE MINIMUM WASN'T GOOD ENOUGH, IT WOULDN'T BE THE MINIMUM: The wisdom of the lazy, the cynical or the enlightened. I leave it to you to decide.

IN ACCORDANCE WITH (IAW): "You'll wear the uniform IAW AFI 36-2903."

INFOCON (Information Condition): These are situational awareness tools to react to threats and attacks on information systems: INFOCON 5 indicates no significant activity and can rise to INFOCON 1, meaning a severe attack has occurred.

JACKED UP: It means the Defenders have taken control of you, and you're probably face down on the ground.

JOCK: Pilot; typically used for fighter pilots (fighter jock).

KIA: Killed In Action.

KNIFE FIGHT: A sarcastic term for a ridiculous bureaucratic spat. And most bureaucratic spats are ridiculous.

KNOCK IT OFF: It means exactly what you think it means, but carries urgency among fliers: It means "Stop everything NOW," because some aspect of the operation has become unsafe or there is an emergency that everyone must be aware of immediately. At this point, everyone stops what they are doing and either returns to base or helps with the situation. Examples: an accident on a range where several aircraft are training; a sudden change in weather or visibility; or an in-flight emergency. In mission briefings, it's common to say, "If XYZ happens, we'll knock it off and talk about it on the ground." As a big airplane guy, I only heard one "knock it off" call: a fighter had crashed on the Red Flag range. The war game stopped so the pilot could be rescued.

LIMFAC (Limiting Factor). LIMFACs are personnel and logistical constraints that impact the planning and execution of military operations.

LMR: Land Mobile Radio. A common means of communication on a base.

LOAD TOAD: An Airman who loads aircraft weapons (missiles, bombs).

MILITARY POWER: Maximum jet engine power without afterburner. In World War II, War Emergency Power was a power setting over 100 percent of the maximum rated power that could be used for a short time, on heavyweight takeoffs or in combat. Today it is called "military power." When these extreme power settings are used, engine life is affected.

MIA: Missing In Action. MIA is a casualty status when a member cannot be accounted for in a combat zone.

MITO: Minimum Interval Take Off. This SAC tactic launches bombers and tankers within seconds of each other, one after another, to get the aircraft into the air rapidly so they aren't destroyed on the ground in a nuclear attack.

MORALE CALL: A phone call home from overseas on a government-provided line at no cost to the member. Typically limited to 10 or 15 minutes once a week. Before the advent of the cell phone and Skype, this was the only way for most overseas deployed personnel to call home.

MUSEUMS: The National Museum of the US Air Force is at Wright-Patterson AFB, Ohio. It is outstanding, with 360 aircraft in its collection and buildings covering 19 acres. Everything from a Wright bicycle to an F-22 Raptor and an MQ-9 Reaper is on display. Highlights include the B-17 *Memphis Belle*; the B-29 *Bockscar*, which dropped the atomic bomb on Nagasaki; the Apollo 15 Command Module, which carried an all-Air Force crew to the moon in 1971; the XB-70 Valkyrie; a collection of Air Force One aircraft; the goblets used by the Doolittle Raiders during their reunions; a Minuteman missile procedures trainer, a TEMPER Tent, EOD equipment and countless artifacts. Viewing it all would take a full day (or more).

The Air Force has many other museums—Dover, Robins, Eglin, Barksdale, Offutt, Travis and Lackland Air Force Bases all have museums associated with them.

MY FUN METER IS PEGGED: "I am not enjoying this."

NCO: Officially, non-commissioned officer. NCOs are the backbone of the American military. In all branches, they ensure our success. Good officers know they succeed because they have good NCOs. Unofficially, it can mean "No Civilian Opportunities." Unfortunately, you occasionally encounter an NCO who stays in the Air Force because he couldn't get another job—not because he likes the service or wants to be an Airman or is good at his work. Don't be this guy.

NCOIC: Non-Commissioned Officer in Charge. The NCO running a shop or an office in its day-to-day tasks.

NKAWTG: "Nobody Kicks Ass Without Tanker Gas." Tankers aren't glamorous; they are flying gas stations. The original purpose of the KC-135 was to support SAC bombers in the Cold War nuclear strike mission. But every air operation since Vietnam has required air refueling to extend the strike force's reach and get it back to base. Tanker crewmembers proudly say:

"Nobody Kicks Ass Without Tanker Gas." Nobody.

NO FACTOR: Not a threat.

NO JOY: Failure to make visual sighting, or inability to establish radio communications; opposite of "Tally ho."

NOTHING IS MORE USELESS THAN RUNWAY BEHIND YOU, ALTITUDE ABOVE YOU AND AIRSPEED YOU DON'T HAVE: This is a simple truth in aviation, but if you're told any of these things in relation to your work, you're failing, and the speaker thinks that you're not worth much to the unit.

NONNER: A term used by fliers and aircraft maintenance personnel referring to those not directly involved in the flying mission. It means "non-sortie generating personnel." Soldiers mutter about the rear echelon, sailors gripe about those who don't go to sea, and Marines disdain everyone who isn't a Marine. On an Air Force flightline, opprobrium is reserved for those who don't directly put airplanes in the air. If your office closes during duty hours "for training," or you aren't out in all kinds of weather, nights, weekends and holidays... well, there's a crew chief on the night shift with a pay or personnel problem, and he's calling you a "nonner." He's only kidding, as far as you know.

OFFENSIVE COUNTER AIR (OCA): Destroying enemy air power through attacks on air bases, destroying aircraft, runways, hangars, supplies and fuel. Destroying assets on the ground and rendering the enemy air force unable to fly is often compared to destroying a hornet's nest rather than each individual hornet. Or, as Curtis LeMay so elegantly said, "Why swat flies when you could destroy the manure pile?"

OIC: Officer in Charge. An OIC is in charge of a section but is not designated a commander. This individual is responsible for overseeing a specific task or operation and ensuring compliance with directives.

OODA LOOP: Col. John Boyd's "Observe Orient Decide Act Loop," a decision matrix. To be inside an opponent's OODA Loop is thinking and moving faster than the opponent, gaining a decisive edge.

O&M: Operations and Maintenance.

OPS: Operations/Operators.

OPLAN: Operations Plan. An OPLAN is a detailed plan for conducting military operations.

OPSEC: Operations Security. Safeguarding mission information is vital.

OPERATION GOLDEN FLOW: The Air Force urinalysis program. People are randomly selected to have someone watch them pee for drug testing. It's said to be a necessity, but I cannot imagine Curtis LeMay putting up with the indignity of what's called a "piss test."

OPPORTUNITY TO EXCEL: A disagreeable job without the time or resources to complete it properly.

ORI: Operational Readiness Inspection. The ORI tests a unit's ability to do its mission under difficult circumstances, simulating a wartime mission.

PASSING GAS: What an aerial tanker does.

PAVE PAWS: A ballistic missile warning system, consisting of a radar about 100 feet tall and its support systems. This was initiated in 1972. There are locations on the Atlantic and Pacific coasts and in Alaska.

PCS (Permanent Change of Station): Reassignment to a different base occurs every few years for most military people, usually based on the needs of the service. Some moves are affected by personal preference. Some people love a place and extend their tours for several years; others count the days till they can leave. The military does a good job of physically moving families and their households. Still, nearly everyone has a horror story, leading to the remark, "Three PCS moves equals a fire for destructiveness." Like most things in life, each person's story is unique, but there are common threads.

PFA (Physical Fitness Assessment): The PFA tests cardiorespiratory fitness, muscular strength, and core endurance. Airmen are assessed on push-ups, sit-ups or cross-leg reverse crunches or timed forearm planks; and a 1.5-mile run or 20-meter Aerobic Shuttle Run or a 2-kilometer walk (if not medically cleared to run). The specifics of these tests and standards often change over time. Some job specialties may have more rigorous testing, such as those in Special Operations. Formerly called the Physical Fitness Test (PFT), and the requirements have changed many times over the years.

PHILIPPINES: The first air unit was assigned there in 1912, under Capt. Frank Lahm. This most distant American colony was a window into Asia for the United States. It was well-positioned to get Americans close to Asian markets and suppliers, though its location put the archipelago squarely in the path of Japanese expansionism in the 1930s. This, of course, led to heavy fighting in the Philippines during World War II. After the war, the US presence in the Philippines was important in the Cold War, Korea and Vietnam. Clark Air Base was the largest USAF overseas base. This lasted until the 1991 eruption of Mount Pinatubo buried Clark AB under two feet of ash. This forced the Air Force to evacuate some 15,000 people in Operation Fiery Vigil, the largest evacuation operation since the fall of South Vietnam in 1975. Clark AB, the oldest USAF overseas base, closed permanently on November 26, 1991, ending more than 90 years of US presence there.

PHYSICAL TRAINING: Working out or playing sports.

PICKLE: To drop a bomb. There's a "pickle button" on the control stick. Pickle probably comes from the pre-World War II idea that good aviators could drop a bomb into a pickle barrel.

PME (Professional Military Education): Continuous learning is required for increasing responsibilities. Captains attend Squadron Officer School (SOS), Majors take Air Command and Staff College (ACSC), and Lt. Colonels and Colonels complete Air War College (AWC). Completion of the appropriate level of PME is a factor in promotion. Enlisted Airmen do Airman Leadership School (ALS) as Senior Airmen to qualify for promotion to Staff Sergeant. Non-Commissioned Officer Academy (NCOA) is required for promotion to master sergeant. Senior Non-Commissioned Officer Academy (SNCOA) prepares Master Sergeants for greater responsibilities and is required for promotion to Senior Master Sergeant.

POCKET ROCKET: The missile badge worn by ICBM missile launch operations and missile maintenance personnel. Ribbons are worn above the pocket, and this badge is worn below the ribbons, on the pocket.

POG: Permanently on the ground. Those who don't fly.

PRACTICE BLEEDING: Training that is more difficult or painful than necessary. We know bleeding is bad. Why would we practice it?

PRESIDENTS: Two US presidents have been Airmen: Ronald Reagan and George W. Bush. Reagan served in WWII when the branch was still the Army Air Forces, and Bush served as a pilot in the Texas Air National Guard.

PROP WASH AND FLIGHT LINE: I couldn't guess how many new, eager, and naïve Airmen have been sent looking for "a gallon of prop wash" or "a hundred feet of flight line." It's a practical joke played on new Airmen.[22]

PUKE: Anyone who flies a different kind of aircraft than you, not a value judgment. "He's a bomber puke, but he's good people."

PUZZLE PALACE: A headquarters. The "Five-sided Puzzle Palace" is the Pentagon.

QUALITY AIR FORCE (QAF): General Merrill McPeak initiated QAF to apply civilian quality management principles to Air Force activities. This was a radical break with the approach that produced outstanding results for years, culminating in the magnificent performance of Airmen in the Gulf War. Don't fix what's not broken.

QUEEP: A task or duty that is unrelated to your primary job or that is unnecessary. For example, firefighters had to complete the training given to

---

[22] Prop wash is the wind generated by an airplane propeller. Flightline is where the planes are parked. No soap or rope required.

office workers on the use of fire extinguishers.[23]

RANDOM ANTITERRORISM MEASURE (RAM): These are security measures that change routine procedures and introduce unpredictability, complicating hostile plans and disrupting surveillance.

RANK HAS ITS PRIVILEGES (RHIP): To some degree, this is true. But if one openly exercises those privileges, it can create resentment. Good leaders keep their privileges to a minimum.

RED FLAG: An aerial combat training exercise at Nellis AFB, Nevada. Since 1975, this highly realistic training for pilots and aircrews has revolutionized preparation for air combat. The antithesis of this might be "Ground Pounder Red Flag," a derisive term used by fliers to describe Professional Military Education, the idea being that several weeks in a classroom environment pushing paper is that experience for non-fliers.

REDUCTION IN FORCE (RIF). Downsizing. Being RIF'd usually depends on AFSC and skill level, evaluations, time in service and staffing in your career field. It is painful, as layoffs always are, but doubly so due to the unique nature of military service. People work at a company; they are in the Air Force. It is an emotional subject. A euphemism for this is "force shaping."

REMOTE: An assignment to a duty station where family members are not authorized to accompany the Airman. Typically, a one-year assignment.

REMOTELY PILOTED AIRCRAFT (RPA) require a rated pilot, sensor or system operator, communications, ground control and support to perform the mission. RPA are expensive but allow air support without exposing aircrew to risks.

RENT-A-CROWD: A big shot visits the base. Who will fill the seats for Senator Schmuckley? The Rent-A-Crowd! It could be an alert team that can't leave the base but can visit the base theater or some other venue. It might be the FTEC (First Term Enlisted Course) Airmen or some other captive group. In any case, a Rent-A-Crowd is those who were told to show up.

REPORT ON INDIVIDUAL PERSONNEL (RIP): The RIP is a master record of one's life in the service: personal data, career information such as dates of service and rank, type of enlistment/appointment, duty stations and assignments, training, qualifications, performance, awards and decorations, disciplinary actions, emergency contacts, etc.

REPORT NO LATER THAN DATE (RNLTD): The date you must be on station at your next assignment. The RNLTD is used to compute travel time and other matters pertinent to the move.

---

[23] https://www.airforcetimes.com/news/your-air-force/2018/06/18/in-war-on-queep-air-force-aims-to-give-squadron-commanders-new-weapons/

ROAD: Retired On Active Duty. This is someone nearing the end of his career who has become less useful, someone serving their time but not being productive. Don't be this guy.

RULES OF ENGAGEMENT (ROE): Combat or contingency operations have rules for how force is used. Airmen are expected to comply with these.

SAN ANTONIO: A military town since the US Army opened a post there in the 1870s. Lt. Benny Foulois was ordered to take the Army's airplane to Fort Sam Houston in 1910 and "teach yourself to fly." From that, the Air Force presence grew into bases named Randolph, Lackland, Kelly, Brooks, Camp Bullis and Medina (now Chapman) Training Annex. The Army kept Fort Sam Houston. To some extent, they are all still there under the unwieldy name Joint Base San Antonio. Randolph is the headquarters of the Air Education and Training Command. Lackland is the home of Air Force Basic Military Training. Kelly Field is adjacent to Lackland and is largely a Reserve operation. Brooks was an Air Force installation from 1918 to 2011; its facilities are now a civilian economic development. Chapman Annex is a training facility for Special Tactics Airmen, and Camp Bullis is an Army training facility. One feature of Randolph AFB that Benny Foulois would recognize is the "Taj Mahal," Randolph's iconic 180-foot-tall water tower, completed in 1931.

SANTA CLAUS: An instructor or evaluator who is generous with grades. But better known, the Air Force tracks Santa Claus. On Christmas Eve 1948, the Air Force issued a press release stating that a sleigh and eight reindeer were picked up on radar "to the north." On December 24, 1955, a department store newspaper ad told kids they could call Santa, but the phone number listed was at Air Defense Command (ADC). The officer in charge took a call from a child and then learned from her mother where she got the number. ADC's Public Affairs office saw the potential for goodwill in this, and the ADC staff was instructed to give callers Santa's "current location." This tradition continues, receiving thousands of phone calls each year. (www.noradsanta.org/en/)

SEAD: Suppression Of Enemy Air Defenses (pronounced "seed").

SECRET SQUIRREL: Slang for classified operations or information. It's taken from an old cartoon featuring a character by that name. If asked about a classified project, "What are you working on?" the quick answer is, "Secret Squirrel stuff." It's also used sarcastically when someone makes more of their actions than is called for. There was also Operation Senior Surprise, which was called "Secret Squirrel" by participants. It was the 35-hour, 14,000-mile sortie by seven B-52s from Barksdale AFB to kick off Desert Storm.

SELECTIVELY MANNED: Personnel in some assignments are carefully chosen due to the complexity, visibility, or importance of the role. These are usually good for one's career advancement.

SENDING A MESSAGE TO HEADQUARTERS: A sarcastic way of saying, "I'm going to the bathroom."

SENIOR ENLISTED LEADER (SEL): These SNCOs provide management, mentoring and career counseling, and address other personnel and logistics issues in a unit.

SENIOR RANKING OFFICER (SRO): In any assembly of military people, the highest ranking person is usually in charge.

SHIRT: Also called "first shirt," this term refers to a First Sergeant. In the frontier Army, the top sergeant had to sign for the junior troops' new gear—such as shirts. Today, the "shirt" is an SNCO who mentors junior members and helps them stay out of trouble. Shirts have been known to help a young Airman who's a bit tipsy and needs a ride home. The first sergeant might tutor the Airmen on best practices in their fields or offer career counseling. Sometimes, the shirt may be the first person to "lay down the law" to an erring Airman. First sergeant is an honored position; a good "first shirt" can make a commander successful. They wear a diamond between their chevrons and the star on their stripes.

SHORT: One is "short" when close to retirement or leaving an assignment. "Sergeant Smith has his orders. He's so short he can't see over a dime."

SILVER BULLET: A module (formerly an Airstream trailer) loaded aboard an aircraft as an enclosed VIP or work area.

SLICK: This can mean a couple of things. It's a standard iron bomb, the kind you've seen falling from B-17s or B-52s in archive footage. They're still around. Slick is also a term for the original cargo-carrying C-130 Hercules[24].

SMOKING HOLE: An airplane crash site.

SNCO: Senior Non-Commissioned Officer; the top three enlisted pay grades E-7, E-8 and E-9.

SOP: Standard Operating Procedure. The way you're expected to do the actions required in your job.

SORTIE: A single mission by one aircraft.

SOUND OF FREEDOM: Jet noise. While it may be a nuisance for those who live near airports to hear aircraft all hours of the day and night, Airmen revel in the sound of jets.

SOUP SANDWICH: "Airman, you've made a real soup sandwich." What could be messier than a soup sandwich? It is an impossible situation with no satisfactory solution. Don't make soup sandwiches.

---

[24] This differentiates the cargo birds from the many specialized C-130 variants.

JET NOISE IS THE SOUND OF FREEDOM

SPACE: On September 1, 1982, the Air Force activated Space Command, which became part of United States Space Command—a joint organization—on September 23, 1985. Space Command's first war was Desert Storm, as the space system infrastructure directly supported forces with communications, navigation, meteorological data and detection of ballistic missile launches.

SPEED: The first US plane to exceed 500 miles per hour was a specially equipped XP-47, reaching 504 miles per hour on August 4, 1944. The XP-80 jet achieved 500 miles per hour soon after that. Most people know Chuck Yeager was the first pilot to exceed Mach 1, in 1947. The first pilot to break Mach 2 was test pilot Scott Crossfield in 1953. The first Mach 3 flight was flown by Capt. Mel Apt. The X-15 took the speed record to its summit, when Maj. Pete Knight flew it to Mach 6.7 (4,520 miles per hour) in 1967.

SPEED IS LIFE: A truth about aviation. Most applicable to fighter pilots in dogfights, and it's true of life: when you slow down, when you stop moving, you're vulnerable and in danger of falling behind, if not out of the sky. Speed is life.

SPEED OF HEAT: Extremely fast.

SPEED OF SMELL: Extremely slow.

SPINS: Special Instructions. Guidance on how to execute a mission.

SPOOLED UP: Jet engines are said to be "spooled up" when they are ready for takeoff. It can also mean a person got fired up about something: "The Chief got all spooled up when he saw Snuffy in that wrinkled uniform."

STEP: Stripes Through Exceptional Performance. Outstanding performers may be promoted under this program to E-5 and E-6.

STRATCOM: US Strategic Command. The successor to Strategic Air Command and the Navy's ballistic missile submarine forces.

TERRAIN AVOIDANCE (TA): The technology that allows an aircraft to stay close to the earth's surface to avoid detection.

TACAN is a radio navigation aid that provides bearing and distance between it and an airplane.

TALLY: Sighting of a target, bandit, bogey or another aircraft; opposite of "No joy." Air Traffic Control: "Cobra 21, you have traffic at 10 o'clock low." Cobra 21: "Tally on the traffic." This comes from the cry "tally ho" when fox hunters have sighted their quarry. Occasionally, fliers will say "Tally ho!"

TANKER TOADS: What people call the KC-135 crew, till they need gas.

TEMPORARY DUTY (TDY): Duty performed away from home station

THE INSPECTOR GENERAL (TIG): IGs—Inspectors General—at all levels conduct assessments of unit readiness, efficiency and discipline. People can file complaints with the IG if they think a commander has wronged them, sort of an official complaints and investigation office. TIG (always with The; like The Ohio State University, it's the official name) – is a three-star general who reports to the Secretary of the Air Force.

THREE DOWN AND LOCKED: A required radio call before landing at an Air Force base, indicating that the landing gear is ready for landing.

THROTTLE BACK: Slow down.

THROW A NICKEL ON THE GRASS: A term of respect, usually applied to those who have "flown west." It is customary to leave a nickel at the grave of a departed flier. There was a Korean War song with the refrain, "Throw a nickel on the grass and save a fighter pilot's ass."[25] There were many such songs written and sung in flying squadrons in that era, as deployed military often had to entertain themselves. Many of these songs are bawdy and are unacceptable in today's Air Force culture.

TLAR: Pronounced "tee-lar," it means "that looks about right." It's an estimate that the job is complete or that the work is okay.

TO ERR IS HUMAN, BUT TO FORGIVE IS NOT COMMAND POLICY: This was a common saying in SAC but is all too often true throughout the Air Force. A no-mistake Air Force will also not learn, because the penalties for mistakes are career-killers.

TOT: Time On/Over Target.

TRANSITION ASSISTANCE PROGRAM (TAP): As people move out of the military and into civilian life, TAP offers help with job searches, networking, resume writing, training and educational opportunities, etc.

TROOP COMMANDER: When Airmen travel in a group, the senior member is designated Troop Commander. There is no pay for this, but someone has to keep track of the passengers and inform the crew of their status.

TS: Top Secret.

UNIVERSAL MANAGEMENT BADGE: Fliers wings, as described by a nonner. This comes from a longtime practice of aviators being in non-aviation leadership jobs despite lacking a background in the field.

VIP CODE: There are many protocol levels: 1 is the President; 2 is the Vice President, Cabinet officers, Governors, Senators, Congressmembers, the Joint

---

[25]There are endless verses of the song, "Save a Fighter Pilot's Ass." These often celebrate the other fliers who might come to a fighter pilot's aid. For example, a tanker crew might join in: "Here's a tanker full of gas to save a fighter pilot's ass, Put your gas hole on the boom and you'll be saved!"

Chiefs of Staff, and former presidents. It goes through level 7: US Consuls, O-6 level officers and justices of the peace.

VOLUNTOLD: If there are no volunteers, somebody will be "voluntold" to do the job. "Why not? You have something else to do, Airman?"

V/R: Very respectfully (closing salutation in correspondence).

WAG: A Wild-Ass Guess, an opinion backed up by nothing. More authoritative is a SWAG (Scientific Wild-Ass Guess) because that means you know a little (but maybe not much) about the topic.

WARHEADS ON FOREHEADS: The Air Force way of saying "striking the enemy."

WARM FUZZY: A feeling of confidence, like a security blanket. "Once you've done this a few times, you'll have a warm fuzzy about doing it on your own."

WASH OUT: To fail a training course. "It was tough when I went through flight training. Almost half of my class washed out."

WEAPONS SCHOOL: In Air Force weapons schools, operators receive advanced tactical training. This was once a fighter-specific program, but now includes every flying weapons system, manned or unmanned, intelligence, space, ICBM and cyber. Graduates are experts and teach their peers. Weapons school graduates are distinguished by a bulls-eye patch on their shoulder, and to be a "patch wearer" is one of the most coveted qualifications in the Air Force.

WHISKEY DELTA: A "weak dude," anyone who just isn't good at his job.

WHITE TOP: Traditionally, the wing commander drives a blue GOV with a white top, rather than blue. In the Strategic Air Command, some aircraft were painted white on the upper fuselage, including those used by headquarters. In that era, Operational Readiness Inspections were "no-notice," meaning a unit might be tasked at any moment to perform its war mission. The only heads-up a commander might have was someone in a control tower calling to tell him, "White top aircraft on final approach." The unit's performance was "make or break" for its commanders' careers. This is depicted vividly in the film A Gathering of Eagles. On the other hand, if a crewmember's wife worked at the Holiday Inn and noticed that someone from Offutt AFB (SAC headquarters) had reserved a large block of rooms, beginning on Monday well, that information might tell a commander that Monday would be the start of a busy and important week.

WHOLE-AIRMAN CONCEPT: One of the problems in any bureaucracy is that it's difficult to evaluate everyone fairly. This concept considers things apart from job performance. Education, community involvement, fitness and such—things that might be more fun than your job. It is both a motivator and a frustration: An Airman wants to be well-rounded, but the one who is a subject matter expert, called on for challenging work and overtime, often feels stymied by the glib, photogenic community ambassador.

WHY NOT MINOT?: Minot AFB, is adjacent to Minot, North Dakota, population 50,000. It's a long way from anywhere. The nearest major urban area is Winnipeg, Manitoba, about five hours away, weather permitting. The summers are hot. The winters are bitterly cold. "Why not Minot?" "Freezin's the reason!" It's not popular with young single Airmen, but families often find it a good assignment for the small-town life, good schools and relatively low cost of living. It is one of two B-52 bases and one of three Minuteman III bases. In SAC, we were told that a young fella should try to go there because "there's a woman behind every tree!" It's on the prairie, very few trees.

WIA: Wounded In Action.

WIC: Weapons Instructor Course. The graduate-level training course focused on the tactical employment of weapons systems.

WINCHESTER: No ordnance remaining. It can be used to refer to specific types of ordnance or all ordnance.

WING KING: Wing Commander. Typically, a colonel or brigadier general, and often the highest ranking person on the base, though some bases have multiple wings, headquarters or missions.

WINGMAN: Second aircraft/pilot in a two-ship formation, one who covers his buddy. Being a wingman is not just for fliers; it is an important part of Air Force culture. It's someone to "cover your six," stand with you, and support and encourage you. Every Airman is a wingman.

WORDS: Further information or directives pertinent to the mission. "We're awaiting words on tonight's mission."

WOWWAJAA: "Without weapons, we are just another airline." This is a reminder that the Air Force's business is deadly serious.

WX: weather.

ZEBRA: A senior non-commissioned officer; they have a lot of stripes.

ZERO-DARK-THIRTY: Operations after midnight/before sunrise. Also said, "Oh-dark-thirty."

ZULU TIME: All military movements and operations are based on Greenwich Mean Time. It's called Zulu because this term (based on the phonetic alphabet—Z is Zulu) would be difficult to mistake for anything else.

CHAPTER 14
BRATS

"I loved being a military brat. I'm eternally grateful to my dad for placing me in this lifestyle. I've never learned so much."
– Wendy Barlow, service brat and Defense Department civilian

Brats are the children of military parents. It's a lovingly bestowed term. The British used the term "British Regiment Attached Traveler" for family members and it's been a US military term of endearment for children since at least World War II. It isn't easy to have a parent deployed, or always preparing to leave, sometimes going to scary places for a long time. And the family moves every few years. Some thrive in this, while others don't enjoy the constant changes. These are some Air Force brats who grew up to be famous:

Ray Allen: The NBA Hall of Famer was born at Castle AFB in California and lived in England, Oklahoma, Germany and South Carolina while his dad was in the Air Force.

John Denver: The singer's dad, Lt. Col. Henry Deutschendorf, flew the B-58.

Pam Grier: This beautiful actress lived in North Carolina, Colorado, Britain and California while her dad was an NCO in the Air Force.

Mia Hamm: The soccer champion's father was an Air Force pilot, and she first played soccer in Italy and then in Wichita Falls, Texas.

Lester Holt: The first black man to anchor a network news program, Holt was born at Hamilton AFB. His father was an NCO.

Swoosie Kurtz: This actress has appeared in many Broadway shows, TV series and movies. She is the daughter of Frank Kurtz, who flew the Boeing B-17D "Swoose" (that's where her unusual name comes from) in World War II.

Annie Leibovitz: The famed photographer became interested in photography while living in the Philippines. Her father retired as a lieutenant colonel.

Mike and Greg Maddux: SMSgt Dave Maddux served from 1957 to 1980, and raised two boys who became major league baseball players.

Mark McEwen: The TV weatherman is the son of Col. Alfred McEwen.

Shawn Michaels: The professional wrestler was born in Arizona, grew up in England and Texas and graduated from Randolph High School on Randolph AFB in San Antonio.

Natalie Morales: A TV reporter and host, Morales was born in Taiwan, where her father was stationed. The family lived in Panama, Brazil and Spain, and she graduated from high school in Delaware.

Priscilla Presley: The actress (and Elvis Presley's bride) was the daughter of an Air Force officer. Priscilla met Elvis in Germany when she was 14.

Robin Roberts: This TV personality's dad was Col. Lawrence Roberts, a Tuskegee Airman who served from 1943 to 1975.

Sen. Jim Webb: The former senator and highly decorated Vietnam War veteran is the son of Col. James Webb, a bomber pilot turned missileer.

Many brats follow a parent into the service. A few you may know:

Lt. Col Michael Anderson: A tanker pilot who became an astronaut, he was born in Plattsburgh, New York, where his father was in aircraft maintenance. He graduated from high school in Spokane, Washington, while his dad was stationed at Fairchild AFB. He was killed on his second Space Shuttle mission in the breakup of the Columbia.

Gen. Benjamin O. Davis Jr.: Davis was the first black general in the Air Force; his father, Brig. Gen. Benjamin O. Davis Sr., was the first black general in the Army.

Gen. David Goldfein: A former CSAF (2016-2020), he is the son of a fighter pilot, Col. William Goldfein.

Lt. Gen. Susan Helms: A pioneering astronaut who flew five Space Shuttle missions, she is the daughter of Lt. Col. Patrick Helms.

Gen. John Jumper: The former CSAF (2001-2005) was the son of Maj. Gen. Jimmy Jumper.

Gen. Michael Ryan: Ryan was CSAF (1997-2001), occupying the same post his father, Gen. John D. Ryan, held during the Vietnam War.

Lt. Gen Richard Scobee: Scobee was an Air Force Academy cadet when his dad, Lt. Col. Dick Scobee, was killed in the Space Shuttle Challenger disaster.

Gen. Janet Wolfenbarger: The first woman to wear four stars in the Air Force, she is the daughter of Maj. Eldon Libby.

CHAPTER 15
MISSION AND WORK

"To be born free is an accident, to live free is a privilege, but to die free is a responsibility. It is why you and I wear these uniforms. To ensure that when we die, we leave a free nation."
- Brig. Gen. "Robbie" Risner

A military service may have a defining victory. For the Marines, it is Iwo Jima; for the Navy, Midway. The Army might claim any number of battles from Saratoga to the Battle of the Bulge. For the Air Force, it's harder to say. Saint-Mihiel, the first air offensive? The Doolittle Raid? Ploesti or Schweinfurt or Berlin? MiG Alley? Linebacker II? Desert Storm? Or maybe the Berlin Airlift and the Cold War that never went hot, in large measure, thanks to Air Force determination and vigilance? The Air Force's roles are countless. Air Force missions go on in war and peace. Here are a few:

AEROMEDICAL EVACUATION (AE) is a lifesaving mission. During World War II, C-46, C-47 and C-54 aircraft were used to transport patients. In Korea, C-47s and C-54s moved patients to Japan or to the US. In 1954, the first dedicated AE platform arrived, the C-131 Samaritan. The C-9 Nightingale (a modified DC-9 airliner) arrived in 1968. Throughout the Vietnam War, tactical airlifters (C-7, C-123, C-130) and helicopters provided battlefield evacuation, while C-9s and C-141s moved patients back to larger medical facilities in Japan and the US.

After the Vietnam War ended, a global AE system with hubs at Rhein-Main AB, Germany, Clark AB, Philippines, and Scott AFB, Illinois, brought military personnel and family members to the US for care unavailable elsewhere. During more recent operations, particularly those in the Middle East, thousands of wounded, sick and injured personnel have been moved by airlifters. When the C-9 left Air Force service in 2005, the KC-135 joined the AE team, with the C-17 and C-130 providing nonstop medical evacuation flights from Afghanistan to Ramstein AB, Germany, a 10-minute drive from the Landstuhl Medical Center, the largest military hospital outside the United States.

AIRBORNE LASER has been studied for more than 40 years. In 1981, a modified

KC-135 aircraft armed with a laser destroyed an aerial target for the first time and, two years later, shot down several air-to-air missiles. In 2002, the YAL-1A, a modified Boeing 747, was used to develop a directed-energy aerial weapon, but these were deemed not operationally viable. Such lasers still seem futuristic.

AIR FORCE ONE (AF1)

Air Force One over Mt. Rushmore

The Air Force flies many high-ranking people, none more important than the president. Any USAF plane carrying the president is designated Air Force One, but people think of the Boeing 747 (Military VC-25). Franklin Roosevelt flew on a Pan American Airways Boeing Clipper and a C-54 to Casablanca, and on a C-54, nicknamed "The Sacred Cow," to Yalta. Harry Truman used that airplane until a C-118 (DC-6) was procured a few years later. Dwight Eisenhower used a VC-121 Super Constellation while he was president, and then John Kennedy was the first president to fly on a jet AF1. This was the VC-137 (a modified Boeing 707), and would serve in the Clinton administration. The current AF1 is expected to be replaced around 2029.

Air Force One is a military aircraft: air refuelable, with global communications capability and electronic countermeasures, enabling the president to perform the duties of the military's commander-in-chief.

A presidential overseas visit is a herculean effort. In November 2006, President George W. Bush went to Singapore, Indonesia, and Vietnam. Air Mobility Command flew 474 airlift and air refueling sorties for this mission, transporting 2,723 passengers and more than 2,400 tons of cargo. This includes the presidential vehicle fleet—typically more than 20 vehicles—and the 10-ton presidential limousines.

Air Force One is a symbol of the United States. Jacqueline Kennedy worked with designer Raymond Loewy to create the elegant steel blue/light blue design we know today. Two VC-137 aircraft were built and retired in 1998 and 2001. One is displayed at the Ronald Reagan Presidential Library, and the other at the National Museum of the US Air Force.

There are two AF1s. Another Boeing 747-type aircraft that may be confused with AF1 is the E-4B, National Airborne Operations Center (NAOC), sometimes called "The Doomsday Plane." It is an airborne emergency command post staffed by more than 100 military members. It can maintain contact and communicate orders globally, even while the United States is under attack.

COMMAND STAFF

The civilian Secretary of the Air Force (SECAF) exercises authority through civilian assistants for acquisition, for manpower and Reserve affairs, for installations, environment and logistics, and for financial management and comptroller.

The Air Staff is led by the Chief of Staff, who is appointed by the president. CSAF is a member of the Joint Chiefs of Staff and a military adviser to the President, the Secretary of Defense, and SECAF. CSAF is responsible for the Air Force's efficiency and readiness, and is accountable for activities assigned to the Air Force by the Secretary of Defense.

Other members of the Air Staff are the vice chief of staff, assistant vice chief of staff, Chief Master Sergeant of the Air Force, five deputy chiefs of staff (manpower and personnel; intelligence, surveillance and reconnaissance; operations, plans and requirements; logistics, installations and mission support; strategic plans and programs), an assistant chief of staff for strategic deterrence and nuclear integration, chief of safety, director of analyses, assessments and lessons learned, judge advocate general, director of test and evaluation, surgeon general, Air Force historian, chief scientist, chief of the Air Force Reserve, director of the ANG, and chief of chaplains.

FIELD ORGANIZATIONS

Ten major commands, the field operating agencies, direct reporting units and their subordinate elements carry out the Air Force mission. In addition, there are the Air Force Reserve and the Air National Guard. Major commands are organized on a functional basis in the United States and a geographic basis overseas. Major commands generally are assigned specific responsibilities. In descending order of command, elements of major commands include numbered Air Forces, Wings, Groups, Squadrons and Flights.

The basic unit for generating and employing combat capability is the Wing. Composite wings operate more than one kind of aircraft and may be configured as self-contained units designated for quick intervention anywhere in the world. Other wings operate a single aircraft type. There are air base and specialized mission wings, such as training, logistics, intelligence and test. Wings are led by colonels or brigadier generals. Within the wing, operations, logistics and support groups led by colonels perform their functions. Groups consist of two or more squadrons, which in turn consist of two or more flights. A squadron is the lowest level of command with a headquarters element. A major or lieutenant colonel leads it and may have 50 to 500 people assigned. Most Airmen are assigned to a flight, a team that does a specific task within a squadron. Aircraft Maintenance Hydraulics, Finance Travel Pay, Personnel Records, Security Forces Operations, etc. It's where you work daily and are expected to become a subject matter expert.

MAJOR COMMANDS

Air Combat Command, Langley AFB, Virginia: ACC provides combat airpower. It trains and sends fighters, bombers in the conventional role, reconnaissance, battle management and electronic combat aircraft to combatant commands. It also provides command, control, communications and intelligence (C3I) systems and conducts global information operations.

Air Education and Training Command, Randolph AFB, Texas: AETC is often called "the First Command" since Airmen start their service there. AETC is responsible for recruiting, technical training, flying training and the continuing education of each generation of Airmen. AETC began as the Air Corps Flying Training Command on January 23, 1942, and became the Army Air Forces Training Command on July 31, 1943. During World War II, Training Command produced about 200,000 pilots, 48,000 navigators, 1.9 million technical training graduates and 2.8 million Basic Military Training graduates. On July 1, 1946, AAF Training Command became Air Training Command (ATC). In 1993, Air University merged with ATC, which then became Air Education and Training Command.

Air Force Global Strike Command, Barksdale AFB, Louisiana: AFGSC activated Aug. 7, 2009. It has 33,000 Airmen and civilians responsible for three intercontinental ballistic missile wings; the Air Force's bomber force to include B-52, B-1 and B-2 wings; the B-21 Raider program; Air Force Nuclear Command, Control and Communications (NC3) systems; and operational and maintenance support to organizations within the nuclear enterprise.

Air Force Matériel Command, Wright-Patterson AFB, Ohio. AFMC was created on July 1, 1992, when Air Force Logistics Command was merged with Air Force Systems Command. AFMC holds over $80 billion in budget authority—about 30 percent of the total Air Force budget—and employs 40 percent of its civilian workforce. The command conducts research, development, testing and evaluation. It provides acquisition and life cycle management services and logistics support for program management of existing and future USAF weapon systems and their components to provide the United States with air power.

Air Force Reserve Command, Robins AFB, Georgia. AFRC has three numbered Air Forces, 34 flying wings, 10 Flying Groups and many other units. AFRC's mission is to provide combat-ready forces to fly, fight and win, and it takes pride in providing about 14 percent of the Total Force for about 4 percent of the manpower budget.

Air Force Special Operations Command, Hurlburt Field, Florida. AFSOC became a major command May 22, 1990. It is the Air Force component of US Special Operations Command, similar to Navy SEALs or Army Green Berets. The Air Force Special Operations team includes combat controllers, tactical air control party, Special Operations weathermen, PJs and all sorts of unique aircraft, doing things such as airborne radio and television broadcasting for psychological operations and assisting the other services' SOCOM elements. AFSOC aircraft include the CV-22B Osprey, AC-130 gunships, EC-130 Commando Solo, MC-130 variants, MQ-9 Reaper, U-28, C-145A and C-146A Wolfhound. Special Tactics troops go through Army Airborne School at Fort Benning, Georgia, Air Force Survival School, Air Force Water Survival Training, Underwater Egress Training, Combat Control School at Pope Field, North Carolina, and Special Tactics Training at Hurlburt Field. They work primarily with Air Force and Army Special Operations Forces but can also be attached to Marine Special Operations and Navy SEAL teams.

Personnel recovery is one of the Air Force's most important missions. Air Force pararescue jumpers (PJs) are trained in search and rescue and combat medicine. PJs are some of the most highly skilled people in the US military: They are extraordinarily fit, trained as combat swimmers and divers, airborne and freefall jump qualified, certified as EMTs and paramedics, and qualified as Special Operations combat troops. It takes about two years to earn the PJ's maroon beret. SERE specialists (Survival, Evasion, Resistance and Escape) integrate survival skills, theater planning activities, and command and control functions supporting personnel recovery operations. Combat controllers call in airstrikes and set up landing zones for Army Special Forces. They specialize in controlling an airfield in covert, austere, hostile, forward or contested environments. CCTs are Combat Control Teams, specializing in controlling an airfield in covert, austere, hostile, forward or contested environments. They establish air traffic control and joint terminal attack control. Becoming a Combat Controller is one of the most difficult training paths in the Air Force, beginning with Air Traffic Controller certification. Special Tactics troops go through Army Airborne School, Air Force Survival School, Air Force Water Survival Training, Combat Diver courses, Free-Fall Parachutist Training, Combat Control School and Special Tactics Training. They work primarily with Air Force and Army Special Operations Forces but can also be attached to Marine and Navy Special Operations teams. Tactical

Air Control Parties (TACPs) work with ground forces to use airpower in support of missions. Combat Weather troops are trained in meteorology to reduce the fog of war through operational weather forecasting and analysis. Weather impacts artillery fire, aircraft employment, force protection and chemical threat response. Combat Weather Airmen are highly trained in tactical combat skills and are airborne and air assault qualified. Special Reconnaissance (SR) Airmen are trained in the use of drones and electronic warfare, and in applying these skills in hostile territory. PJs, Combat Rescue Officers and SERE members conduct personnel recovery in both hostile and peacetime environments.

Search and Rescue (SAR) is how personnel are recovered in emergencies—aircraft crashes, training mishaps in the field, etc. Combat Search And Rescue (CSAR) is the recovery of personnel in denied and hostile environments. SAR is peacetime search and rescue. The leaders are Combat Rescue Officers (CRO), part of the Special Tactics Officer (STO) team.

One unique tradition in the Special Warfare community is Green Feet. You may see them stenciled on a flight suit, an aircraft, or tattooed on someone's posterior. It comes from the "Jolly Green Giant" HH-3E helicopter, which left footprint-like indentations when it landed in grassy fields.

Air Mobility Command, Scott AFB, Illinois: Activated June 1, 1992, AMC traces its history to the establishment of the Air Corps Ferrying Command on May 29, 1941. As the air component of the US Transportation Command, AMC's Total Force executes Global Mobility and Global Reach, able to respond anywhere in the world in a matter of hours. This is accomplished through AMC's four core mission areas: Airlift, Air Refueling, Air Mobility Support, and Aeromedical Evacuation. AMC has approximately 110,000 Total Force personnel and operates the Air Force Expeditionary Center, Eighteenth Air Force, and the 618th Air Operations Center, along with ten installations.

Pacific Air Forces, Hickam AFB, Hawaii. PACAF has 46,000 military and civilian personnel serving primarily in Hawaii, Alaska, Japan, Guam and the Republic of Korea. Approximately 320 fighter and attack aircraft are assigned to the command. PACAF has three numbered Air Forces (5th Air Force in Japan, 7th Air Force in Korea, and 11th Air Force in Alaska). PACAF began as Far East Air Forces (FEAF), on August 3, 1944, at Brisbane, Australia.

US Air Forces in Europe/Africa, Ramstein AB, Germany. USAFE-AFAFRICA directs Air Force units from the United Kingdom to Turkey, as well as air and space operations in Europe, western Asia and most of Africa. It is the oldest continuously active USAF major command, activated initially in January 1942, as the 8th Air Force was established. On August 7, 1945, it was designated as United States Air Forces in Europe (USAFE). Today, there are 35,000 active duty and civilian employees assigned.

Field operating agencies and direct reporting units report directly to

Headquarters US Air Force. They are assigned specialized missions that do not fit any specific command. Examples: The Air Force Personnel Center is a field operating agency. The Air Force Academy is a Direct Reporting Unit.

AIRLIFT

Billy Mitchell envisioned aircraft dropping supplies and paratroops, and a few supply drops occurred in World War I. In the 1920s, cargo and passenger aircraft began to be produced. In 1932, the 1st Air Transport Group began flying supplies and equipment from depots to where they were needed.

Before the US entered World War II, the Air Corps Ferrying Command was created to deliver lend-lease aircraft overseas. From early 1941, the new command was to move aircraft from factories to aerial ports or operational bases and to provide air transport as required. Ferrying Command was under Brig. Gen. Robert Olds (father of fighter ace Robin Olds).

Air routes were developed from CONUS to Britain, to Alaska, the Soviet Union, to Australia, and the Southwest Pacific via Hawaii. A south Atlantic route tied the US to West Africa, via the Caribbean, then from Brazil to Senegal, and on to Morocco and Europe, or the Mediterranean, or to the Middle East, India and China. Later, a Mid-Atlantic route was established through the Azores to Europe and North Africa.

On September 17, 1941, during Army exercises, DC-3s dropped paratroopers. DC-3s were the civilian version of the C-47, an aircraft that would deliver cargo and troops for decades.

Soon, personnel and priority cargo were traveling worldwide by air. Daily flights to remote places became routine. Commercial airlines supported the war effort in a unified Air Transport Command. Air transport became an essential function. The airlines had many planes, trained pilots and navigators, and extensive experience operating on schedules across far-flung routes. More than half of the civilian airliner fleet entered service, and many personnel joined the AAF or worked as contract crewmembers.

Hap Arnold realized that unified control of air transport was essential to maximize the use of assets. In June 1942, Ferrying Command became Air Transport Command, and a Troop Carrier Command (for parachute cargo drops and paratroop insertion) was created, dividing strategic and tactical airlift. In March 1942, General Olds had a heart attack (a heart condition would take his life in 1943), and he was replaced by Brigadier General Harold L. George, who led ATC until the end of the war.

Air Transport Command expanded from 37,000 men to 200,000 (mostly overseas) and from 130 to 3,000 aircraft. At first, the DC-3/C-47 was the primary airlifter. By war's end, four-engine aircraft such as the C-54 and the

C-87 Liberator Express (a transport version of the B-24) were shouldering the load. American Airlines President C. R. Smith was commissioned as a colonel in ATC, and he expanded operations quickly. Smith put ATC in charge of the Hump airlift operation from India to China. In July 1945, ATC carried 275,000 passengers and 100,000 tons of mail and cargo, nearly all of it overseas.

When the Air Force was established in 1947, ATC became the Military Air Transport Service (MATS), incorporating Navy air transport units. This organization would oversee military air cargo and passenger movement, as well as Special Air Missions (SAM) that carry the president and senior government officials. In 1966, MATS was replaced by Military Airlift Command (MAC). Navy units that supported MATS were reassigned as fleet logistics support squadrons. Transcontinental air travel became common after World War II. Aircraft developed during the war, such as the Douglas DC-4 (C-54) and the Lockheed Constellation (C-69), could cross the Atlantic Ocean with a fuel stop in Newfoundland or Ireland. The first plane to cross the Atlantic with more than 100 people aboard was an Air Force C-74 Globemaster on November 18, 1949, flying from the US to England.

MAC's primary mission was strategic airlift, similar to that of Air Transport Command. C-124s, C-141s and, later, C-5s and C-17s were its backbone.

The Vietnam War saw a steady flow of cargo and passengers across the Pacific. Operation Blue Light was a peak, lifting 4,600 tons of equipment and 3,000 troops from Hickam AFB, Hawaii, to Vietnam, from December 27, 1965, to January 22, 1966. The C-141 Starlifter served extensively during Vietnam and was an effective AE aircraft. C-141s flew to Hanoi in 1973 to bring home returning POWs.

Operation Nickel Grass was the resupply of Israel during the Yom Kippur War in 1973. MAC C-141 and C-5 transports carried gear from tanks to airplane parts to Israel. America's European allies refused to allow MAC flights bound for Israel to refuel on their soil. Fortunately, Portugal allowed Lajes Field in the Azores to be used. Air refuelable airlifters would alleviate this problem. This prompted a program to equip the C-141 with air refueling capability.

In 1983 and 1989, operations in Grenada and Panama put MAC at the forefront as the US moved swiftly to take down dictators in the Western Hemisphere. When Iraq invaded Kuwait in 1990, President George H.W. Bush sent American armed forces to Saudi Arabia. The 82nd Airborne Division was airlifted from the US to Saudi Arabia in a few days. C-130s were deployed to Saudi Arabia to support the arriving ground forces.

As the Cold War ended, the Air Force reorganized. On June 1, 1992, MAC became AMC—the Air Mobility Command, the Air Force component of the United States Transportation Command (USTRANSCOM). In addition to airlift, the new organization added air refueling to its mission, as KC-135 and

KC-10 tankers were shifted from SAC. Air Mobility Command is responsible for global air mobility, supporting combat operations, deployment and humanitarian aid. AMC conducts airlift, aerial refueling, Aeromedical Evacuation and Special Assignment Airlift Missions (SAAM), such as Presidential Support.

Aircraft assigned to AMC are airlifters: C-5 Galaxy, C-17 Globemaster III, C-130 Hercules; tankers: KC-135 Stratotanker, KC-46 Pegasus; and personnel transport: C-40 Clipper, C-37 Gulfstream V, C-21 Learjet. The 89th Airlift Wing moves the president and senior officials using VC-25, C-32, and "business jet" aircraft. AMC can be augmented by the Civil Reserve Air Fleet (CRAF) if the president activates it. CRAF aircraft are civilian commercial planes designated to support military forces in a crisis.

The backbone of modern airlift has been strategic airlifters, nicknamed "T-tails" for their distinctive empennage (tail assembly), and the ubiquitous C-130 Hercules. The most versatile aircraft ever built, the 130 has been adapted in countless ways. It has gone to the polar ends of the earth as a ski-equipped bird and has caught film cartridges dropped from satellites. It has flown reconnaissance missions and controlled drones. The -130 is an airborne command post, an electronic warfare platform, and some can broadcast TV and radio signals to carry out psychological operations. Herks perform aerial spraying, refuel helicopters and participate in search and rescue operations. They fly into the eyes of hurricanes to do weather research and gather data to forecast storm tracks. Hurricane Hunters originated in 1944, and the Herk has been used in this mission since the 1960s. Airmen still fly into hurricanes to protect America. They've even served as bombers, dropping the 15,000-pound Daisy Cutter and MOAB (Mother of All Bombs). Still, the Herk isn't much of a dogfighter, and no fighter model has appeared—yet! More than 70 years later, the C-130 remains the best tactical airlifter, and in its newest form, the C-130J, it will continue to be so. On February 17, 1999, the first C-130J, with its six-bladed propellers, advanced avionics and shorter takeoff and landing capability, arrived at Keesler AFB, Mississippi. The C-130J offers significantly improved performance with no flight engineer or navigator, saving training dollars.

The **LOCKHEED C-141 STARLIFTER** was the Air Force's primary strategic airlifter for decades. The first operational C-141aircraft was delivered to Travis AFB on April 23, 1965. The Air Force bought 284 of them. It could carry up to 70,000 pounds of payload, including 154 troops, 123 paratroopers, or combinations of troops and supplies. In 1966, a Starlifter became the first jet to land in Antarctica. The plane was a workhorse, delivering troops and equipment to Southeast Asia. In 1977, the C-141 was "stretched" by 23 feet, adding air refueling capability. That same year, the aircraft made its first transatlantic crossing without a navigator, using a dual Inertial Navigation System, similar to that being used on airliners at the time. C-141s dropped paratroops in Panama and delivered thousands of tons of cargo and nearly

100,000 passengers during Desert Shield and Desert Storm. The C-141 flew the mission to the Pegasus runway near the McMurdo Research Station in Antarctica until 2005. In 2006, the C-141 flew its last mission to a combat zone, a medevac sortie to Balad AB, Iraq. On May 5, 2007, a C-141 made its final passenger flights at Wright-Patterson AFB. The passengers were 125 former POWs. This aircraft was nicknamed the "Hanoi Taxi" because it had transported the first group of Vietnam prisoners of war from Hanoi on February 12, 1973. The next day, the aircraft joined the National Museum of the Air Force collection.

The **LOCKHEED C-5 GALAXY** was designed to meet the Air Force's requirement to move outsized and oversized cargo. With its front and rear access doors, the cargo compartment is large enough to accommodate a tank. The C-5 can carry up to 36 cargo pallets or almost limitless combinations, such as two Abrams tanks, seven Huey helicopters, up to 281,000 pounds. It can carry all of the Army's air-transportable combat equipment. The cargo compartment is 121 feet long—one foot longer than the Wright brothers' first flight! There are 35,000 cubic feet of storage in the cargo bay (five times that of a C-141A; by comparison, a 53-foot over-the-road trailer has about 4,000 cubic feet of space), and the C-5 can carry up to 270 passengers. On June 6, 1970, Gen. Jack J. Catton, commander of MAC, accepted delivery of the first C–5 Galaxy. The C-5 had a rocky start. Its reliability was poor; its complex hydraulics and landing gear were designed to allow the plane to land on dirt strips and to "kneel" to enable drive-on, drive-off cargo movement. A fortune has been spent on modernization and newer versions of the C-5. It is uniquely capable.

The first of 222 **BOEING C-17A GLOBEMASTER IIIS** arrived at Charleston AFB, S.C., in June 1993. Capable of delivering outsized cargo in a tactical environment, the C-17 reduces cargo handling by reaching airfields usually only accessible to C-130s. This reduced the number of convoys at risk on potentially hostile roads between outposts and bases in Iraq and Afghanistan. It does phenomenal work: In 2006, one C–17 airdropped 32,400 pounds of humanitarian goods to four locations in central and eastern Afghanistan in just 40 minutes. On January 30, 2011, three C–17 aircrews dropped 120 bundles, each loaded with four 55-gallon fuel drums. More than 26,000 gallons of fuel meant providing a 30-day supply. One airdrop on October 11, 2007, got 62 bundles of cargo, weighing more than 85,000 pounds, to coalition forces in remote Paktika province in southeastern Afghanistan. They released their cargo in less than three minutes using highly accurate delivery procedures that minimized exposure to hostile fire. Airlifters moving cargo over hostile territory meant fewer truck convoys transporting supplies where there was a high risk of attack.

On August 31, 2006, the joint precision airdrop system (JPADS) was first used on a combat mission when a C–130 airdropped supplies to a US Army unit in Afghanistan. The JPADS is a satellite-guided parachute system that allows cargo to be dropped accurately from altitudes up to 25,000 feet, beyond the range of enemy ground fire. By mid-2007, JPADS enabled Airmen to make high-altitude airdrops as close as 10 to 15 yards of the intended point of impact.

On the subject of Airlift, everyone knows something about what pilots do, but loadmasters are the experts in putting passengers, cargo and paratroops on an airplane so that it can be offloaded—or airdropped—safely and efficiently. They fly on C-130s, C-17s and C-5s. Every cargo load is a balancing act. The loadmaster must ensure the aircraft maintains a safe weight and balance. The placement of a heavy object—trucks, tanks, pallets of cargo—is a matter of calculation. If you're dropping it out of the back of the airlift aircraft, then that's another issue. A rapid change in weight and balance can be fatal if not done properly. Loadmasters are often junior crew members, but their skills are respected and counted on in every flight.

AIR FORCE OFFICE OF SPECIAL INVESTIGATIONS (AFOSI, often just called OSI) is the investigative arm of the Air Force. Thanks to TV, *NCIS* (Naval Criminal Investigative Service)[26] is more famous, but the two are similar. OSI handles counterintelligence, counterespionage, counterterrorism, illegal technology transfer, computer infiltration and criminal investigations. OSI investigates felony crimes, black marketing and financial crimes such as fraud in contracting, in handling military funds and property, and in pay and allowances.

AIR AND SPACE OPERATIONS CENTER (AOC). An AOC (pronounced "AYE-ock") is a type of command center. Many are established regionally—Alaska, Korea, Japan, Europe, and Qatar; others perform a functional role—Strategic, Cyber, and Transportation. JAOC (pronounced Jay-ock) refers to a Joint (multi-service) AOC, run by a JFACC (Joint Forces Air Component Commander), while CAOC (pronounced Kay-ock) refers to a Combined (multi-national) AOC, run by a CFACC (Combined Forces Air Component Commander). The AOC generates the Air Tasking Order (ATO), the air battle plan for a day or a stated period.

AREA OF RESPONSIBILITY (AOR): An AOR defines areas where combatant commanders are tasked with preparing for combat or humanitarian operations. The CENTCOM AOR is probably best known since Central Command (the Middle East) was the focus of the Global War on Terrorism. But there is also Africa, Europe, Indo-Pacific, Southern (South America), Northern (North America) and Space. Functional combatant commands exist: Special Operations, Cyber, Strategic and Transportation.

ASAT: Anti-Satellite weapons have been tested. On September 13, 1985, a missile launched from an F–15 destroyed a satellite orbiting at 17,500 mph, 290 miles above Earth. While an enemy's satellites might seem an obvious target, there are still questions about the utility of such technologies.

---

[26] Mark Harmon, the star of *NCIS*, is the son of football star Tom Harmon, an Air Force fighter pilot in World War II.

ASTRONAUTS

It's no surprise that the Air Force has sent about 100 astronauts into space. Becoming an astronaut is a rare honor, the culmination of intense effort, work, education and training.

The Air Force played a significant role in Project Mercury. Three of the original seven astronauts were Airmen: Captains L. Gordon Cooper Jr., Virgil I. "Gus" Grissom, and Major Donald K. "Deke" Slayton. Cooper had flown fighters and was a test pilot at Edwards AFB. He flew the final Mercury mission and was on Gemini 5 before retiring in 1970. Grissom served in World War II, earned an engineering degree at Purdue, and returned to the Air Force. In Korea, Grissom flew F-86s, then moved into flight test. He flew the second suborbital Mercury mission and then Gemini 3. He died in the launchpad fire of Apollo 1. Deke Slayton flew B-25s in the Mediterranean and A-26s against Japan during World War II, then became a test pilot. He was disqualified from space flight due to a heart condition but was cleared to fly the Apollo-Soyuz mission in 1975.

The Air Force's Gemini astronauts were a mix of familiar Mercury names and new guys. Deke Slayton was the director of operations, and Gordon Cooper and Gus Grissom flew on Gemini 3 and 5, respectively. On June 7, 1965, Gemini 4, with Majors James A. McDivitt and Edward H. White, set a US endurance record of 97 hours, 30 seconds in space. Major White became the first US astronaut to "walk" in space during this mission. Major Thomas Stafford was the pilot for Gemini 9, Mike Collins flew Gemini 10, and Buzz Aldrin flew Gemini 12. Two Airmen died while training as astronauts: Charles Bassett and Ted Freeman, both killed in T-38 crashes. The Air Force recommended Capt. Ed Dwight to NASA. A fighter pilot with an aeronautical engineering degree, Dwight would have been the first black astronaut. However, NASA did not accept him. He did eventually go into space with Blue Origin in 2024.

Apollo saw more Airmen in space. Four Airmen have walked on the moon: Buzz Aldrin (Apollo 11, 1969); David Scott, James Irwin (Apollo 15, 1971); and Charles Duke (Apollo 16, 1972). Eight Airmen have been to the moon without landing: Frank Borman, William Anders (Apollo 8, 1968); Thomas Stafford (Apollo 10, 1969); Michael Collins (Apollo 11, 1969); Jack Swigert, Fred Haise (Apollo 13, 1970); Stuart Roosa (Apollo 14, 1971); Al Worden (Apollo 15, 1971).

One Apollo crew was all-Air Force: Apollo 15 (July 26—August 7, 1971) was flown by Col. David R. Scott, Lt. Col. James B. Irwin and Maj. Alfred M. Worden. The mission was the first to use a lunar rover vehicle on the surface of the moon. Scott and Irwin spent almost three full days on the moon.

Since Apollo, astronauts have not been as well-known to the public. Most still come from flying fighters and flight test, but astronauts are some of the most accomplished people imaginable. While about 100 Airmen have been astronauts, some of the best known Air Force astronauts have been:

John Blaha: five Space Shuttle missions. He flew 361 combat missions in fighters during the Vietnam War.

Guion Bluford: three Space Shuttle missions. The first African American to go into space, he was an F-4 pilot with a PhD in Laser Physics.

Karol Bobko: three Space Shuttle missions. The first USAFA graduate in space, he flew fighters and was a test pilot before joining NASA.

Roy Bridges: two Space Shuttle missions. Bridges was a Misty FAC with 262 combat sorties in Vietnam, then was a test pilot. He retired as a major general.

Curtis Brown: six Space Shuttle missions. He flew the A-10 before becoming a test pilot and an astronaut.

Kevin Chilton: three Space Shuttle missions. He flew fighters, then went into flight test and NASA. He then resumed his Air Force duties, rising to four stars and completing his career as commander of Strategic Command.

Eileen Collins: four Space Shuttle missions. Collins was the first woman to pilot the Space Shuttle. She was a C-141 pilot, then a test pilot.

Dick Covey: four Space Shuttle missions. He flew 339 missions in fighters in Vietnam, earning five DFCs. After that, he went into flight test, then joined NASA. He was a Capsule Communicator for six missions, and it is his voice in the audio of the Challenger disaster, saying "Go at throttle up."

Brian Duffy: four Space Shuttle missions. A fighter pilot—he flew in the same wing with Kevin Chilton—and a test pilot before becoming an astronaut.

Joe Engle commanded two Space Shuttle missions, including the program's second orbital flight. He also flew Shuttle Approach and Landing tests. Engle wore astronaut wings before he went to NASA, earned while flying the X-15.

John Fabian: two Space Shuttle missions. Fabian was a KC-135 pilot, flew 90 tanker missions in Southeast Asia, and was a professor of aeronautics at the Air Force Academy. He operated the Shuttle's robotic arm.

Gordon Fullerton: two Space Shuttle missions. Fullerton was a member of the Air Force's Manned Orbiting Laboratory (MOL) project, after flying F-86s and B-47s. After the MOL's cancellation, he worked in support of the Apollo program and flew Shuttle Approach and Landing tests.

Fred Gregory: three Space Shuttle missions. Gregory flew rescue helicopters in Vietnam and earned three DFCs in 550 missions. He then moved into the Astronaut Office at NASA.

Henry Hartsfield: three Space Shuttle missions. He was a fighter pilot, a test pilot, and in the MOL program. The last seven years of his USAF career were at NASA.

Susan Helms: five Space Shuttle missions. Helms was an aerospace engineer in the Air Force when she was chosen to be an astronaut. She went to the International Space Station (ISS), logging 210 days in space. Resuming her Air Force career, she served in Space Command and Strategic Command. She was promoted to lieutenant general and was Fourteenth Air Force commander.

Thomas Jones: four Space Shuttle missions. A B-52 aircraft commander who left the Air Force to earn a doctorate in Planetary Science, he joined NASA after his USAF service.

Steven Lindsey: five Space Shuttle missions. He flew the RF-4 and became a test pilot. His missions took him to the International Space Station.

Pam Melroy: three Space Shuttle missions. She was a KC-10 pilot with combat experience in Desert Storm, then became a test pilot in the C-17 program.

Mike Mullane: three Space Shuttle missions. Mullane was a WSO in the RF-4 in Vietnam, earning a DFC. He became a flight test engineer, then was assigned to NASA from 1978 to 1990. After retirement, he wrote one of the most candid (and hilarious) memoirs ever published by an astronaut, *Riding Rockets: The Outrageous Tales of a Space Shuttle Astronaut*.

Ellison Onizuka and Dick Scobee were the Air Force members aboard the Space Shuttle Challenger in 1986, both on their second space mission. Onizuka was an engineer throughout his career. Scobee had enlisted in the Air Force, earned a degree and a commission, and flew combat missions in Vietnam before entering flight test.

William Pogue flew on Skylab 4, an 84-day mission from November 1973 to February 1974. He flew 43 combat missions in Korea and was a Thunderbird pilot who taught mathematics at the Air Force Academy and completed the British Empire Test Pilot School. He was expected to fly in the Apollo program before it was curtailed.

Charles Precourt: four Space Shuttle missions. He flew fighters, flight test, and then on to NASA. His fluency in Russian made him an ideal candidate to fly the 1998 mission in which the Shuttle docked with the Russian spacecraft Mir.

Jerry Ross: seven Space Shuttle missions, more than any other Airman. He was an engineer in the Air Force and joined the space program in 1979. His space missions took place between 1985 and 2002, during which he performed nine extravehicular activities.

Ron Sega: two Space Shuttle missions. Sega, an Air Force Academy graduate, was an instructor pilot and taught physics at the Academy. Moving to the Reserve, he earned a PhD in electrical engineering and taught at the University of Colorado. He became an astronaut in 1990. After his space flights, he returned to academia. He was Assistant Secretary of Defense for Research and Engineering from 2001 to 2006. He retired as a major general in the Air Force Reserve.

Brewster Shaw: three Space Shuttle missions. Shaw flew the F-100 and F-4 during two Vietnam combat tours, winning eight DFCs. He became a test pilot and, in 1978, joined NASA.

Lacy Veach: two Space Shuttle missions. An Air Force Academy graduate, he flew 275 missions in Vietnam, earning three DFCs and a Purple Heart. He was a Thunderbird in 1976-77 and transferred to the Texas Air National Guard in 1981 while working at NASA. He was named an astronaut in 1985. He died of cancer at 51 in 1995.

Some space superlatives:

Farthest from earth: Navy Capt. Jim Lovell, former Air Guardsmen Jack Swigert and Fred Haise traveled farther from Earth than anyone else on Apollo 13: 248,655 miles from home.

Fastest humans: The Apollo 10 crew achieved a speed of 24,791 miles per hour, approximately Mach 32, the fastest speed attained by humans so far. The crew was USAF Col. Thomas Stafford and Navy Capts. John Young and Gene Cernan.

Most orbital flights: Jerry Ross is one of three people to launch into space seven times. (John Young and Franklin Chang-Diaz are the others)

FLIGHT PAY: One of the benefits of military aviation is flight pay. It began in July 1914, in a time when at least half of the early military fliers died in accidents. As late as 1933, 4 percent of the pilots died in that year alone. Although the accident and fatality rate is significantly lower, flight pay is an incentive to keep military pilots in uniform, particularly when airlines are hiring. In December 2003, the Secretary of the Air Force issued a policy letter authorizing aviation career incentive pay and flying credit for pilots of unmanned combat aerial vehicles.

HUMANITARIAN AND DISASTER RELIEF

Since the Berlin Airlift, lifesaving relief has been a hallmark of the Air Force. But that was not the first time that aviation was used to bring assistance to people suffering disasters. In 1919, Air Service planes from Kelly Field, Texas, dropped food supplies to flood victims along the Rio Grande. Aerial resupply was a frequent use of pre-WWII aircraft. In December 1935, the

5th Bombardment Group bombed the Mauna Loa volcano, diverting its lava flow away from Hilo, Hawaii, an event reflected on the 23rd Bomb Squadron's patch.

During World War II, Airmen flew humanitarian missions everywhere. In September 1944 the AAF hauled food to France, where the roads and railroads had been wrecked. In the last week of the war, "Operation Chowhound" brought food to Dutch civilians.

It would take a book to list all of the Air Force's humanitarian missions. A few examples will provide a picture of the scope and volume of these operations.

August 1988: As civil war in southern Sudan created thousands of refugees, the 60th and 436th Military Airlift Wings delivered more than 70 tons of plastic sheeting, shelters, food and other supplies. USAF C-5s moved 500 Canadian United Nations peacekeepers to Turkey and Iraq. As wildfires burned millions of acres in Yellowstone National Park, Active, Guard and Reserve aircraft transported 4,000 firefighters and 2,000 tons of firefighting equipment. At the same time, C-130s with modular airborne firefighting systems (MAFFS) were deployed to drop fire retardant chemicals to fight the fires. A C-141 delivered a 200-bed emergency hospital, and another flew 29 tons of medical supplies to Sao Tome off the West Coast of Africa. That's just one month, and not unusual for the Air Force.

In August 1992, after Hurricane Andrew hit Florida, the Air Force delivered more than 21,000 tons of relief supplies and transported more than 13,500 passengers. In September 1992, after Typhoon Omar hit Guam, the USAF transported 750 relief workers and 2,000 tons of supplies to the island. Typhoon Iniki hit Kauai in September 1992, and USAF units airlifted more than 12,000 passengers, evacuees, and relief workers.

A major humanitarian tasking came during Hurricane Katrina. WC–130 aircraft tracked this hurricane as early as August 23, 2005. It first made landfall in Florida and then roared across the Gulf of Mexico, hitting the coast near New Orleans on August 29. The storm ravaged New Orleans, causing $100 billion in damage, and killing nearly 2,000 people. By August 31, US Northern Command activated JTF-Katrina at Camp Shelby, Mississippi. Air Force helicopters flew 648 sorties, 599 of which were search-and-rescue missions that rescued 4,322 people. Air Force fixed-wing aircraft flew 4,095 sorties, including 3,398 air mobility missions, evacuating nearly 27,000 displaced persons, plus 2,600 aeromedical evacuations. More than 11,450 tons of relief supplies were brought in. Air Force medics at the New Orleans International Airport treated nearly 17,000 patients. The ANG's efforts included 4,132 sorties, moving 34,639 relief personnel and evacuating 2,000 people. Over 20,000 people were rescued by ANG helicopters, and about 11,000 tons of cargo were moved by ANG airlifters.

Some unique humanitarian missions include delivery of $15 million in medicine and medical supplies to Bishkek, Kyrgyz Republic, in September 2004; an airlift to the Republic of Georgia in August 2008 moved 78 tons of humanitarian aid and medical supplies. A massive effort to assist Japan following a magnitude 9.0 earthquake in March 2011; the quake led to a tsunami and wrecked a nuclear powerplant. The effort, named Operation Tomodachi, involved assets from all the US armed services. C-17s delivered search and rescue equipment and personnel the day after the quake, and nearly 200 tons of relief supplies and equipment were delivered to Japan.

ISR (INTELLIGENCE, SURVEILLANCE AND RECONNAISSANCE): This aspect of the Air Force ranges from satellites gathering data and monitoring potentially hostile actors, to quiet professionals listening to communications among those actors. The public is most familiar with aircraft such as the SR-71 and the U-2, but the RC-135 stands as a silent warrior in this mission, in peace and war, as do the thousands who do unsung work for the nation in less-visible roles. In 1966, the Air Force received its first SR-71 unit at Beale AFB, California. The Blackbird remains an incredible aircraft: It could fly faster than Mach 3, and beyond 80,000 feet. It required special fuel and maintenance support and could go from New York to London in just one hour and fifty-four minutes, and from London to Los Angeles in three hours and forty-seven minutes. Its highest reported speed was 2,193 miles per hour, and its greatest reported altitude was 85,000 feet. Those might be the most amazing facts about this aircraft. Or the most amazing thing is what it might have done that remains classified. Or that this is 1960s technology!

The first U-2 high-altitude, long-range reconnaissance aircraft was delivered in 1957. Ideal for overflight reconnaissance missions, the U-2 was little-known until May 1960, when Francis Gary Powers was shot down by a SAM deep in the Soviet Union. Powers spent about 21 months in a Soviet prison before he was exchanged for captured Soviet spy Rudolf Abel. Killed in a helicopter crash in 1977, Powers was awarded a posthumous Silver Star for his service, a joint operation of the Air Force and the Central Intelligence Agency.

The U-2's finest hour came during the Cuban Missile Crisis, finding Soviet weapons and revealing the extent of Soviet weapons buildups in Cuba. One U-2 was shot down and the pilot, Major Rudolf Anderson, was killed. The U-2 is still operational, with more built in the 1980s. It conducts all-weather, day-and-night missions above 70,000 feet.

LAW OF ARMED CONFLICT (LOAC): American forces will respect international law. The law of armed conflict demands humane treatment of enemy military and civilians in US hands, causing no unnecessary destruction or suffering. American forces do all they can to mitigate the harmful effects of war.

MAVERICK: With apologies to Top Gun, the Air Force had a Maverick in the air long before Hollywood. On December 18, 1969, the Air Force tested the Maverick, an air-to-surface television-guided missile capable of attacking moving targets at short range. Designated the AGM–65, it is carried by many fighter and attack aircraft, including the MQ-9 Reaper. In 1971, Hughes Aircraft was awarded a $70 million contract to build 2,000 Maverick (AGM–65A) air-to-surface missiles. Mavericks are still in use, and about 70,000 have been built and used in wars from Vietnam to the present.

OPERATION DEEP FREEZE: The Air Force has a long history of polar flight. On May 3, 1952, a ski-equipped C-47 made the first successful North Pole landing. A few years later, a Navy version of a C-47 landed at the South Pole, part of many years of Navy support for the United States Antarctic Program. The Air Force took over the Antarctic resupply mission in the 1990s. Strategic airlifters fly into McMurdo, and ski-equipped C-130s carry passengers and cargo to the South Pole. On December 20, 2006, C-17 crews completed an airdrop mission to the South Pole, delivering 70,000 pounds of supplies to the National Science Foundation team at the South Pole, a cargo load similar to that of four air-land C-130 missions.

PEOPLE IN THE AIR FORCE:

USAF Strength 2024 vs. 1989 numbers:

	1989	2024
Active Duty	571,000	325.000
Officers	102,000	61,000
Pilots	20,338	13,000
CSOs	8,550 (Navigators)	3,500
Air Battle Managers	2,182	1,900
RPA Pilots	0	2,600
Female	74,000	70,000
DAF Civilians	262,000	167,000
Reserve	84,800	100,000
ANG	116,000	106,000

Peak strengths: In World War II, the AAF topped out at 2.37 million in 1944. In the Korean War, the USAF's maximum strength was 983,000 in 1952. During Vietnam, USAF strength peaked at 904,000 in 1968. Its post-Vietnam maximum was 608,000 in 1986.

The minimum USAF strength since its founding in 1947 was 311,000 in 2015. USAF has been over 500,000 for most of its existence, but has been under 400,000 active members since 1996.

Most Common Duty Areas, Officer	1989	2024
Operations	39,000	29,000
Intelligence	3,329	5,200
Nurse	5,295	3,200
JAG	1,319	2,300
Cyber Operations	0	3,000
Force Support	3,000	2,400

Most Common Duty Areas, Enlisted	1989	2024
Aircraft Maintenance	146,000	87,000
Security Forces	39,000	37,000
Medical	28,000	27,000
Intelligence	13,000	21,000
Enlisted Aviators	9,000	13,500

PRISONER OF WAR (POW): Many Airmen are captured over enemy territory. Aviators fly deep into enemy territory, far from friendly forces. Despite the superhuman efforts of the Air Force's CSAR forces, Airmen are sometimes taken prisoner. Hunger, maltreatment and fear have been constants for prisoners of war.

In World War II, a downed Airman's first risk was the vengeful mob. There are many stories of Airmen being lynched by hostile civilians. One former POW said that he was glad to see German soldiers arrive when he landed because they protected him from a howling throng of civilians. If the German military captured an American flier, he was likely to live through the war: 98 percent of the American POWs in Germany survived. It wasn't easy. Food was, at best, monotonous and of poor quality, and then scarce as the war turned against Germany. The Germans generally permitted Red Cross food parcels to supplement prisoner diets.

Though most American fliers captured by the German military received correct treatment, there were glaring exceptions. SSgt. Bernard Scharf was the son of a German immigrant to the United States and a gunner in the 91st Bomb Group. He was shot down on his first mission. He hooked up with the French Resistance but, along with other evaders, was betrayed and turned over to the Gestapo. The Nazis interned them at Buchenwald Concentration Camp. One day, Luftwaffe Col. Hannes Trautloft visited. He had heard rumors of mistreatment of Allied fliers, and under the guise of viewing bombing damage, he visited the camp. While there, a man in the striped uniform of an inmate called out to him, in German, that he was an American flier. Trautloft pulled rank on the SS commander and spoke with the man, SSgt. Scharf. The

American explained that there were more than 160 Allied Airmen held there in horrible conditions; several men had died, and they feared the Gestapo would kill them. Trautloft arranged their transfer to a Luftwaffe-run "Stalag." They survived the war. Other Airmen were not so fortunate. Under Hitler's orders, some Airmen were sent to concentration camps, and American Airmen are known to have died in Buchenwald, Dachau and Mauthausen.

Those captured by Japan faced an enemy who ignored any standards regarding the treatment of human beings, much less POWs. Downed Airmen were often murdered on the spot by military or civilians. Those who were taken prisoner were still in danger. A US Army website describes a 700-calorie daily diet coupled with 12-hour days of hard labor. Nearly 40 percent of US prisoners held by Japan would die from starvation, slave labor and mistreatment. The Louis Zamperini story in the movie and book *Unbroken* provides a glimpse of that. Many POWs were transported and died in "Hell Ships," held in cargo holds, in tropical heat, near-darkness, with minimal food, water and sanitary facilities. Many of these unmarked vessels were also freighters carrying war supplies to Japan and were sunk by American submarines.

The war in Korea brought another 7,000 Americans into enemy prisons. About 40 percent of these would die there. The psychological and physical torture of prisoners, along with deliberate efforts to turn prisoners against their country and to make pro-Communist statements (brainwashing), was shocking.

The Vietnam War saw about 771 known Americans in enemy custody, and about 115 died there. These prisoners were some of the longest-held in American history. Hayden Lockhart was the third American and the first Airman captured in North Vietnam. He was flying an F-100 when his plane was hit by ground fire, and he had to eject over North Vietnam. He evaded capture for ten days but was caught on March 2, 1965. He was 26 years old, his wife was seven months pregnant, and he would spend 2,905 days in enemy hands. He was released on February 12, 1973. The treatment of prisoners by the enemy was appalling. Torture and brutality were common. In a war that was virulently opposed by so many, the heroism of the POWs was widely recognized.

The fate of the MIAs (Missing In Action) was a lingering wound for Americans. Widespread suspicion that the Vietnamese and their allies—the Soviets and Communist Chinese—were withholding some of these Americans was never alleviated. The fact is, a lot of people are lost in war—at sea, in deep, distant jungles and deserts and, unfortunately, entire bodies are destroyed as a result of explosion or fire. Could Americans have been carried away into a Soviet Gulag, or was their technical expertise used by our enemies? We may never know. Recent wars have seen few Americans taken by the enemy. Being a prisoner of war is a terrifying aspect of the military experience.

Becoming a prisoner of war is one of the most traumatic events anyone could experience. John Giraudo, Wendall Phillips, Richard Keirn and Vernon Ligon lived through this nightmare twice.

John Giraudo: Like many of his generation, Giraudo volunteered after Pearl Harbor. He was a 20-year-old captain assigned as a B-24 pilot in Italy. He was shot down over Regensburg, Germany, in February 1944. He spent 14 months as a prisoner of war. Continuing his military career, he flew F-86s in Korea. On June 16, 1953, he was shot down over North Korea while strafing trucks and was missing in action. That he survived was unknown until he was released in October 1953. Giraudo flew 100 missions in F-105s in Vietnam and retired as a major general in 1977.

Wendall Phillips: He was captured for the first time in 1944, when the C-47 transport on which he was a radio operator went down in Belgium, and German soldiers found him. He spent 33 days in German hands before walking out through a gap in a fence. He worked his way back to Allied lines and, in 1945, returned to action in China. There, he survived another crash and was taken prisoner by the Japanese. He was liberated at the end of the war after enduring brutal captivity.

Richard Keirn: He was a 20-year-old B-17 pilot on his eighth mission when his plane was destroyed on September 11, 1944. He was wounded but survived as a POW. He remained in the Ohio Air National Guard after the war and returned to active duty in 1956, flying jet fighters, until he was downed by a surface-to-air missile (SAM) over North Vietnam on July 24, 1965, while flying an F-4C. This time, Keirn spent nearly eight years as a POW, returning home in 1973.

Vernon Ligon: On April 22, 1944, Ligon was flying his 35th mission in P-47s when he was shot down over Belgium. He spent the next year in German POW camps until his liberation in April 1945. He was a career Air Force officer, serving in staff positions and flying the B-47 bomber. On November 19, 1967, he was on a reconnaissance mission over North Vietnam, flying an RF-4C, when he was shot down. He was released from North Vietnam in March 1973.

German POW interrogator Hanns Scharff was the exception to the brutal torturer battering prisoners to obtain information. His father was a Prussian officer killed in World War I, and his mother was from a wealthy family. Scharff was married to a South African woman and was fluent in English. He was drafted in 1941 and assigned to interrogate POWs. Scharff was in his mid-30s and took a friendly approach to the American and British fliers who came through his office. With his background, he knew British customs and slang and quickly picked up American ways. He "befriended" prisoners, bringing them homemade food, taking them sightseeing and introducing them to Luftwaffe pilots, which often led them to talk freely. He built a "roster" of fliers so that sometimes, when a prisoner walked into his office, there was familiar information—even photographs—about himself, his commander and his squadron mates. Often, prisoners assumed he knew more than he did and unwittingly filled in the gaps.

Maybe you have seen Hanns Scharff's postwar work. After World War II, he emigrated to the United States and worked as a mosaic artist. His work adorns universities, office buildings, private homes, the California State Capitol and Cinderella's Castle at Walt Disney World. It is a small world, after all.

TROUBLE

The Uniform Code of Military Justice (UCMJ) is federal law governing members of the uniformed services. This legal code combines criminal offenses (such as murder or drunk driving) with specifically military offenses (cowardice, failure to follow orders, absence without leave, etc.). You could face jail time in the military for being late to work (unlikely, but possible). You can't just quit the job whenever you want. Being court-martialed is as serious as any other criminal proceeding. But most offenses don't lead to criminal charges, so there is an array of administrative punishments that can be levied on those who don't behave according to the military's rules.

Unit commanders usually handle these administrative tools, but supervisors may initiate them. They begin with an LOC—Letter of Counseling. This is a corrective action appropriate for shortcomings not criminal or illegal, but that affect job performance, morale and discipline. Perhaps tardiness in reporting for duty, or failure to participate in or accomplish a job task for which one has been trained. Then comes an LOA—Letter of Admonishment. This is an administrative punishment for violating standards, more severe than a LOC. Counseling implies that we believe you have the potential to do better. Admonition says you had better improve.

The Letter of Reprimand (LOR) is more severe than an LOA. An LOR leads to an Unfavorable Information File (UIF), affecting appraisals, promotion recommendations and awards nominations. Administrative penalties can include denial of pass privileges, extra duty or training and administrative demotions. The UIF includes punishment under Article 15, court-martial convictions, civilian court convictions and control roster action (substandard work performance).

The next step is Article 15. The UCMJ's Article 15 covers non-judicial punishments that may be offered instead of trial by court-martial, typically for less serious violations. It's usually used when administrative measures have not led to corrected behavior. The commander reviews the evidence, decides, and has a spectrum of non-judicial punishments that may be imposed. The accused has all the rights of a defendant in court, and the government must prove the alleged offense beyond a reasonable doubt. The accused may appeal if found guilty or may request a court-martial. The possible consequences?

If convicted by court-martial, this federal conviction is on the record for the rest of one's life. Article 15 punishment is serious, but it's not a criminal conviction. The offender may be fined or demoted, lose future promotion or training opportunities, or be placed in short-term correctional custody and have a UIF created. An Article 15 stays in personnel records and can hinder reenlistment. For Senior NCOs and officers, it is a roadblock to further advancement. However, outside the military, Article 15 punishment is unlikely to be an issue. Within the military, if the offender "keeps his nose clean," some of the punishments may be suspended or reduced.

Beyond these actions, there is the court-martial system, with juries, judges, evidence, procedures and punishments.

A few other administrative terms you might come across are:

Memorandum For Record: The MFR is an informal document to record and preserve information.

LOA: Letter of Appreciation. That's a good thing!

LOE: Letter of Evaluation. LOEs are evaluations for brief periods, typically under the supervision of someone they don't usually work for—on a deployment, for example. They can be used to highlight good (or bad) performance so that home station leadership is aware of it.

TROUBLE WITH A CAPITAL T

Like any organization, the Air Force makes mistakes, missteps or fails. Some screwups are matters of opinion: such as changing the dress uniform in 1994 or changing the officer rank insignia on that new uniform. Or the leather nametag with rank on the BDU at that time. Or doing away with the three major flying commands. And the unpopular Airman Battle Uniform.

The Defense Department budgeting process is impossible to understand. Billions of dollars are often unaccounted for, and procurement programs drag on, and costs balloon. Sometimes, Airmen complain that it appears budget decisions are made using Ouija boards, coin flips or "rock, paper, scissors." Ideally, it would be based on what America needs. In reality, it's more a matter of what America can afford. On April 6, 2009, Defense Secretary Robert Gates announced sweeping program cuts to reduce acquisition of the F–22 and the C–17. Also cancelled were plans for a second airborne laser aircraft, satellite programs, and the next-generation bomber was delayed. I don't know who is at fault —if anyone —in all this. But politicians and their partisans clamor for military involvement and support for allies. You cannot say, "We must assist

Ukraine, or stand up to China, or..." if you do not have the strength to prevail in an armed conflict arising from these positions.

But unfortunately, in any human endeavor more serious screwups occur. The Transcontinental Air Race in October 1919 saw 63 Air Service airplanes try to fly a New York – San Francisco route and back, about 5,400 miles. Half the planes did not complete the trip and seven men died.

Crashing the prototype B-17 in 1935 because someone left the control locks in the elevators was a fatal mistake. Parking the planes in neat rows in the Philippines and Hawaii on December 7, 1941, was a bad move. Operation Cobra saw the Air Force bomb American troops in 1944, killing more than 100. Air Force generals went along with Rolling Thunder, a failure of leadership. In one week in 1994, Fairchild AFB, in Washington, lost a B-52 in a crash and suffered a mass shooting. There were missed signs about a reckless pilot and a disturbed airman.[27] The horrendous shootdown of two Black Hawk helicopters in Iraq in 1994 killed 26 people. A tragic mishap in 1996 that killed Secretary of Commerce Ronald Brown and 34 others when a CT-43 airplane crashed into a hillside near Dubrovnik, Croatia.

Friendly fire in combat is a terrible thing. In Desert Storm, eleven US Marines and nine British soldiers were killed by A-10s. An F-16 pilot killed four Canadian soldiers, and a B-52 killed three American troops, among the more serious incidents in Afghanistan. In Iraq, US Marines and soldiers were killed by Airmen, in terrible mistakes of misidentification.

There have been spies, traitors and untrustworthy Airmen: Martin Monti defected to fight for Germany in 1944. Brian Regan, Edward Buchanan, Jeffrey Carney, Christopher Cooke, Allen Davies, Francisco Mira, Frank Nesbitt, Bruce Ott, Ronald Wolf and Monica Witt all spied against the United States.

Losing a nuclear weapon would be, at the least, a screwup. The most famous incident occurred near Palomares, Spain in January 1966. A B-52 bomber and a KC-135 tanker collided during air refueling and crashed near Palomares, killing seven of the 11 crew members. In this case, all four nuclear bombs were recovered, one only after an extensive search in water 2800 feet deep. But in other cases, the bombs were never recovered:

Feb. 13, 1950—A B-36 with a nuclear weapon on board developed severe engine problems over the North Pacific Ocean. They dropped the A-bomb into the ocean to lose weight and stay airborne.

March 10, 1956—A B-47 vanished over the Mediterranean Sea. Nothing of the crew, plane or weapons has ever been found.

Feb. 5, 1958—A B-47 and an F-86 collided in flight near Georgia's Tybee Island. The bomber jettisoned a nuke into the Atlantic.

---

[27] *Warnings Unheeded* by Andy Brown is an excellent account of these tragedies.

Jan. 24, 1961—A B-52 caught fire over North Carolina and exploded. The bomber was carrying two nuclear bombs. One was found, but the other was not. Its tail was found, and the weapon is believed to lie more than 150 feet below the surface.

In all these cases, there is much speculation and conflicting stories.

In August 2007, six nuclear-equipped missiles were mistakenly transported from Minot AFB, North Dakota, to Barksdale AFB, Louisiana, aboard a B–52. There, they were left without the security measures that safeguard nuclear weapons. The Air Force investigated, and several people were punished.

On March 24, 2008, the Air Force announced that it had mistakenly shipped ICBM parts to Taiwan in 2006. Taiwan's government reported the mistake immediately, but the USAF took over a year to account for the shipment. Six generals and nine colonels were punished in the wake of this.

Then, in July 2008, three members of a missile crew were discharged after falling asleep while holding classified materials. The three were in a secure crew rest area, and the outdated launch codes remained in their locked containers. These stayed with the crew in a facility guarded by heavily armed Security Forces. Perfection is the standard required when handling classified information. Think about this when you hear of political figures mishandling classified materials.

These incidents led to significant changes in leadership. Defense Secretary Robert Gates requested the resignations of CSAF Gen. T. Michael Moseley and Air Force Secretary Michael Wynne.

The buffoonery did not stop there. On April 27, 2009, a VC–25A from the presidential fleet caused an incident in New York when it flew low over the city for photographs taken from an F–16. While the Federal Aviation Administration knew what was happening, no one informed the public.

Then there's the Expeditionary Combat Support System (ECSS). It was expected to streamline USAF logistics operations with a single logistical system. After spending a billion dollars, the Air Force admitted the system had not "yielded any significant military capability." Your tax dollars at work.

Is it a screwup when the Air Force doesn't see that a paradigm has changed? World War II bombers needed fighters for protection in enemy airspace, itself a hard-won lesson that defied a whole lot of pre-WWII airpower thought. Bombers did not have escort fighters capable of reaching deep into Germany until 1944, when the P-51 arrived.

However, no fighter plane could carry out a global bomber escort mission. In the 1950s, an attempt was made to have the B-36 carry an escort fighter along, and a recovery apparatus was developed. Experiments included hooking a fighter to the bomber's wing or placing a trapeze-like device in the bomb bay and carrying the fighter under the bomber's fuselage. Similar methods carried various experimental aircraft to different altitudes, so it wasn't foolish. Bombers had lifted the X-1, the X-15 and others. Designers soon discovered that a fighter plane sacrificed performance, and the maneuvers to hook up were risky, so bombers' defensive tools came to be electronic countermeasures and tail-mounted guns.

The modern world may face changes such as cheap drones and other unconventional weapons that counter expensive weapons systems. Our task will be to prepare for all the possibilities.

The Air Force is tremendously effective and committed to service. But like all human enterprises, it occasionally falls short of expectations. Airmen work to do better.

CHAPTER 16
FIRSTS AND LASTS, SUPERLATIVES

"I'm a citizen of the United States of America and I'm no second-class citizen either... This is my country and I believe in her, and I will serve her, and I'll contribute to her welfare whenever and however I can. If she has any ills, I'll stand by her until in God's given time, through her wisdom and her consideration for the welfare of the entire nation, she will put them right."
- Chappie James

African American firsts: The Air Force takes pride in the service of all Americans. Before Harry Truman signed the order to integrate the armed services, the Air Force was moving on the issue. Here are a few pioneers:

First black pilots: the Tuskegee Airmen, in 1942.

First black Air Traffic Controller: Alex Boudreaux, 1946

First black general: Benjamin O. Davis Jr.

First black four-star: Daniel "Chappie" James. A Tuskegee Airman, James flew fighters in Korea and Vietnam. He received his fourth star and became commander of NORAD and Air Defense Command on September 1, 1975. Since then, Bernard Randolph, Lloyd "Fig" Newton, Benjamin O. Davis, Lester Lyles, Larry Spencer, Ed Rice, Darren McDew, Charles Brown, Anthony Cotton and Randall Reed have achieved this

General Daniel "Chappie" James was a pioneering fighter pilot

rank. The Air Force takes pride in the fact that everyone has a chance to go as far as their talents and skills will take them.

First black Chief Master Sergeant of the Air Force: CMSAF Thomas N. Barnes was an aircraft maintainer and flight engineer and served in Korea and Vietnam. He was named CMSAF from 1973 to 1977. Today, the Crew Chief of the Year award is named for him.

First black Chief of Staff (CSAF): Charles Q. Brown. General Brown was an F-16 pilot who became CSAF in 2020; in 2023 he became chairman of the Joint Chiefs of Staff.

First black female Chief Master Sergeant: Dorothy Holmes. She enlisted in 1949 and became the first woman to have a 30-year career in the Air Force, retiring in 1979.

First black Cadet Wing Commander of the United States Air Force Academy: Cadet Edward A. Rice Jr. of Yellow Springs, Ohio, on August 3, 1977. Rice would wear four stars by the time he retired in 2013.

First black astronaut in space: Lt. Col. Guion S. Bluford, aboard Challenger on August 20, 1983. He was an F-4 pilot in the Vietnam War.

The first black woman to reach Air Force general rank: Marcelite Jordan Harris, who pinned on a star on September 3, 1990. In 1971, she was the first female aircraft maintenance officer. She retired as Air Force Director of Maintenance.

Lt. Col. Stayce D. Harris became the first black woman to command a USAF flying squadron, the 729th Airlift Squadron at March AFB, California, on February 24, 2001. Harris became assistant Vice Chief of Staff of the Air Force, retiring as a lieutenant general.

The first minority generals in the Air Force were Clarence L. Tinker and C.V. Haynes. Tinker was born in Indian Territory, later the state of Oklahoma, in the Osage Nation. He was a brigadier general before the war, and after Pearl Harbor, he was placed in command of Hawaiian air defenses. As the Japanese fleet retreated from Midway, Tinker led a flight of B-24s against them but his plane was seen to crash into the Pacific. No bodies were recovered. C.V. Haynes was a grandson of Chang Bunker, one of the 19th-century conjoined "Siamese" twins. Vance joined the Army in 1917, became one of the first B-17 pilots, and flew the lead bomber in the interception of the *Rex* in 1938. He led bombers under Claire Chennault in China. He retired as a major general.

First Chief Master Sergeant of the Air Force: Paul W. Airey was the first CMSAF. He was a radio operator on B-24s in Europe, shot down on his 28th mission. He spent 10 months as a POW. He was a radio maintainer and a first sergeant throughout his career. Airey served two years as CMSAF and retired in 1970.

Last WWII veteran CMSAF: Richard Kisling, who served in the Army in the European Campaign.

Last Korean War veteran CMSAF: Thomas Barnes. He was a flight engineer and maintainer. Later CMSAFs served during the Korean War, but not in-theater.

Last Vietnam veteran CMSAF: Eric Benken, an administrative specialist.

Last Chief of Air Service without wings: Charles Menoher, who left the Air Service in 1921.

First Chief of Air Service who earned wings: Mason Patrick. He was also the oldest man to qualify as a pilot at age 59 in 1922.

First Chief of Staff with no combat experience: Lew Allen. He had flown B-50s and B-36s in SAC, then earned a Ph.D in nuclear physics.

Last World War II veteran to be CSAF: Charles Gabriel (1982-1986). He entered West Point in 1946 and thus qualified for the World War II Victory Medal. Gabriel was also the last Korean War veteran Chief of Staff; he shot down two MiGs in Korea.

Last World War II combat veteran to be CSAF: George S. Brown (1973-1974) flew B-24s and was on the Ploesti raid.

Last Vietnam War vet to be CSAF: John Jumper (2001-2005) flew 1,400 hours in combat in C-7s and F-4s.

First navigator promoted to four-star general: Bernard Randolph (1987). To date, he is the only navigator to reach this rank.

First non-rated four-star: William F. "Bozo" McKee. General McKee was a West Point graduate who was a Coast Artillery officer before World War II. Assigned to duty with AAF units, he transferred into the Air Force after the war. He was a skilled administrator who rose to be the Vice Chief of Staff of the Air Force, and after retirement, head of the Federal Aviation Administration.

First Reserve four-star: Jimmy Doolittle (an honorary promotion in 1987); Maryanne Miller was the first serving Reservist to be a four-star general in 2018. She commanded Air Mobility Command.

General Bernard Randolph: navigator and space systems expert.

First ANG four-star: Craig McKinley, 2008.

First USAFA grad promoted to brigadier general: Harold W Todd, 1978.

First USAFA grad four-star: Gen Hansford T. Johnson, USAF, 1989.

First USAFA grad CSAF: Gen Ronald R. Fogleman 1994.

Most combat missions flown: It's hard to say. Among fighter pilots, Col. Ralph Parr flew 641 combat missions and Col. Harold Snow flew 666, over World War II, Korea and Vietnam. C-123 and A-37 pilot Col. Ralph Dresser flew 750 missions. Helicopter pilot Col. James Fleming flew 810 missions in Vietnam. Lt. Gen. Thomas Richards flew 624 missions in Vietnam in B-52s, O-1s, T-28s and other aircraft as a Raven FAC. C-7 and B-52 pilot Lt. Col. Phillip Anderson flew 1,000 sorties in the Caribou in Vietnam—28 in one day. Col. Win DePoorter flew 657 missions in A-1s. MSgt. Hewitt Dunn flew 104 bomber missions in World War II, and 60 more in Korea.

Enlisted pilots: About 3,000 enlisted pilots served between 1912 and 1957; 157 died in the line of duty. Many of the enlisted pilots flew liaison or support aircraft, that is, smaller or non-combat planes.

The first enlisted pilot was Corporal Vernon Burge. Burge enlisted in 1907, was one of the first aircraft mechanics, and in 1912 got his pilot certificate while stationed in the Philippines. He served until 1942, retiring as a colonel, with 4,000 flight hours.

The last enlisted pilot was MSgt. George Holmes. A sailor in World War I, he switched to the Army in 1919. Holmes earned his pilot wings in 1921 but chose to stay enlisted because of officer force reductions. He served in many roles—observation, flight test, bombers and Air Mail pilot. During World War II, he was commissioned and flew B-25s in the Pacific. He reverted to master sergeant as the services returned to peacetime strength. Holmes retired in 1957, with 9,000 flight hours.

The longest serving Airman was MSgt. Horst Tittel. Born in 1884 in Germany, Tittle came to the United States in 1904. He enlisted in the Army in 1908 and transferred into the Aviation Section of the Signal Corps in 1917. A master sergeant when America entered World War I, he was commissioned and returned from France as a captain. He reverted to master sergeant in 1919. He served in that rank until 1942, when he was commissioned again during World War II. He served in North Africa and Italy, rising to lieutenant colonel. He chose a reversion to master sergeant in 1945[28] and retired from the Air Force on September 30, 1959, with 51 years of service. He died in 1975.

---

[28] Why would he choose to "revert" to enlisted status? Many officers commissioned in wartime had reserve commissions, meaning they were not guaranteed a job in the full-time military force after the war and the drawdown that followed. Tittel and others chose to complete their military careers as sergeants, but would retire in the highest rank they had held. So Master Sergeant Tittel retired as Lieutenant Colonel Tittel.

Last Medal of Honor recipients: Americans noted the passing of the last World War II Medal of Honor recipient, Hershel "Woody" Williams, in 2022. The last World War II Medal of Honor Airman was Jay Zeamer, who died in Maine on May 11, 2007, at age 88. Eddie Rickenbacker was the only World War I Medal of Honor Airman to survive the war; he died in 1973. All four Air Force Medal of Honor recipients in the Korean War were killed in action. The last surviving Airman recipient from the Vietnam War is Col. James Fleming. The only Airman to receive the Medal of Honor in the GWOT was TSgt. John Chapman, killed in action at Takhur Ghar, Afghanistan.

The last living American Ace of World War I was Arthur Raymond Brooks, who died in 1991 at age 95. He was an MIT graduate who joined the Air Service in 1917 and was credited with six aerial victories. After the war, he worked for Bell Labs, testing air navigation and communications systems. His SPAD XIII aircraft is on display at the National Air and Space Museum.

The last living American Army Air Forces Ace of World War II was James L. McCubbin, who died in 2025 at age 103. A student at the University of Missouri when America entered the war, he flew P-51 Mustangs in Europe. He was credited with four victories but a paperwork error that missed his last two combat victories left him short of the coveted status of an "ace." He went into engineering work after his military service. At the time of his passing, the paperwork to obtain official credit for the fifth and sixth German aircraft he destroyed was making its way through the bureaucracy. The American Fighter Aces Association was convinced of his case, and added him to their roster.

Last World War II pilot on active duty: Col. Thomas J. Tredici. Born in 1922, he enlisted in 1942 and flew B-17s with the Eighth Air Force, logging 18 combat missions from November 1944 to the end of the war. After World War II, he went to college, then medical school, graduated in 1952 and returned to the USAF. He served as an eye doctor and flight surgeon, including service in the Vietnam War. Upon his return from Vietnam, he became Chief of the Aerospace Ophthalmology Branch at the USAF School of Aerospace Medicine, Brooks AFB, Texas, from 1966 till he retired at age 65 in 1987.

Following his military retirement, Dr. Tredici continued as a Senior Scientist and Clinical Function Chief of the Aerospace Ophthalmology Branch at the School of Aerospace Medicine, working in civil service until 2011. He continued working as an Emeritus Scientist and consultant until he died in 2021.

The passing into history of the World War II generation leaves me saddened. I was fortunate to know many of these veterans. They did a great and noble thing to preserve freedom. They were my heroes.

The biggest Air Force base is Eglin AFB, with almost half a million acres in Florida. But there are vast, desolate ranges in the US where the military conducts training. The largest are in Nevada, Arizona and Utah. Each of these is similar in size to Yellowstone National Park and is surrounded by vast airspace. The Alaska Training Range is 65,000 square miles, about the size of Wisconsin.

The Air Force base with the most Airmen is Lackland AFB, Texas, with about 16,000 Airmen and 6,000 civilians assigned. Wright-Patterson AFB, Ohio, has 15,000 civilians and about 7,000 military personnel.

The Air Force's longest serving aircraft is the B-52, in service since 1955. The C-130 has been in the inventory since 1956. The KC-135 has been in service since 1957, as has the U-2. These have been modified extensively, but an old-timer would recognize them. Curiously, all these types are more than 60 years old, but all were flying less than 60 years after Kitty Hawk.

THE GRATEFUL DEAD AND THE AIR FORCE: In the 1940s, there was a young man named John Parber from Arizona. John met a girl, and one thing led to another. Then she left town in a hurry, never telling John why. She was pregnant, and it was the 1940s. She had the baby in San Francisco. A family in an affluent suburb adopted the baby boy, and the girl moved on with life. John joined the Air Force and became a bomber pilot. He flew B-29s in the Korean War and B-52s in the Vietnam War, rising to colonel and having a family. The baby he never knew about was named Bob Weir, and became a member of the Grateful Dead 1970s rock band.

One day, a woman called the Grateful Dead's office (even rock bands have staff), claiming to be Bob's mother. Her story checked out, and she talked with Bob, revealing his biological dad's name. Bob called a private investigator, and it turned out that Bob's dad was Col. John Parber, the commander at Hamilton AFB, just north of San Francisco. Bob hesitated to contact the man but eventually he picked up the phone. He introduced himself, and the colonel replied, "The only Robert Weir I know is the guy who plays with the Grateful Dead." The colonel's sons—Bob's half brothers—were Grateful Dead fans. Bob and Colonel Parber met and learned they had an astonishing amount in common, even similar personalities.

The Parbers and Bob Weir's family became close. One of Bob's half brothers had been a professional musician but died young. Bob was given one of his guitars, which he played on stage with the Grateful Dead. Col. John Parber died at age 89 in 2015.

THE VENTURES AND THE AIR FORCE. The 1960s rock group's first drummer was a kid named George Babbitt, who was too young to play the nightclubs where they were getting noticed. He went to college, joined AFROTC, and became a general. The Ventures were inducted into the Rock & Roll Hall of Fame. At a concert in Dayton in 1998, the Ventures invited the four-star to sit in on the drums with them. Google "George Babbitt Ventures." The grin on the general's face when he hits the drum solo is worth a look!

MURPHY'S LAW ("what can go wrong, will go wrong") was invented by an Airman, Maj. Edward Murphy, in 1949. He was an engineer working with JP Stapp's rocket sleds, then on cockpit design and crew escape systems.

PUBLIC PERCEPTIONS OF THE AIR FORCE

Aviation captured people's imagination from the first days of flight. People stared in wonder at flying machines, things that were once thought impossible. In 1984, my great-grandmother reminisced about the first time she saw an airplane in flight, in 1921. Our generation might have the same wonder at virtual reality or artificial intelligence.

The changing perceptions of airpower are evident in those the public knew as heroes: Eddie Rickenbacker and other aces of the First World War. Billy Mitchell, with his headline-grabbing antics in the 1920s. There were countless heroes in World War II newsreels. Fighter aces were popular, bomber crewmen were in the news, and men like Dick Bong, Bob Morgan, Don Gentile and John Godfrey were all well known in the 1940s. Cargo crews had a moment in the spotlight during the Berlin Airlift—Gail Halvorsen represented the kindly American, offering aid to his former enemies while dropping candy to the former enemy's kids. Fighter pilots recaptured the spotlight in Korea—Joe McConnell, Jim Jabara and Pete Fernandez were featured in the media as they raced to become America's top jet ace. Hoyt Vandenberg, the boyish, good-looking Air Force Chief of Staff, was often featured in the media. Test pilots such as Chuck Yeager, Mel Apt, Iven Kincheloe and Pete Everest made headlines as jets went higher and faster and began to approach space. The deaths of Apt and Kincheloe provided a stern reminder that this was a deadly business. As the Soviet threat loomed, grim-visaged Curtis LeMay became America's best known Airman, and his instantly ready, powerful and modern SAC forces stood prepared for Armageddon.

There hasn't been an Airman who has captured the public's imagination since then. Astronauts are often mentioned in news reports, but their military status is downplayed; they are explorers who come in peace. Their military service isn't as noteworthy as their status as star voyagers. Vietnam War POWs were briefly lionized, but few Americans could name any of them today (although within the military, they are legendary). Victorious fighter pilots and heroic bomber and transport crewmembers are seldom noted in the news. Reconnaissance crews work in the anonymity of secrecy, and missile combat crews quietly tend their terrible weapons across the sparsely populated western plains.

The Thunderbirds provide one of the best known public faces of the Air Force. They gave their first performance on June 8, 1953. Other demonstration teams flew in the 1930s, but this team existed to garner publicity and public awareness of the Air Force. More than 300 million people have seen the

Thunderbirds in 82 countries. Their equipment has been F-84s, F-100s, F-105s (briefly), F-4s, T-38s and, since 1983, F-16s.

Airmen and now Guardians place satellites in orbit, maneuver them and enable a nation to reap the benefits of space-borne communications, weather observation, reconnaissance, navigation and countless other applications. Thousands of Airmen have gone to war since 9/11, most of them little noticed, though valiant special tactics forces occasionally gain recognition. Still, their role is often literally in shadow and darkness, unknown to the American public. Today, aviation is almost taken for granted, and for that, the Air Force deserves great credit.

CHAPTER 17
A CHANGING WORLD

"I'm an Air Force officer first, a pilot second, and then Nicole. The female part is last... My job is to be the best right wingman that I can be."
- Lt. Col. Nicole Malachowski, first female Thunderbird pilot

WOMEN IN THE AIR FORCE: Women have been in the Air Force since it was the Army Air Forces. As America mobilized for World War II, a separate unit, the Women's Army Corps (WAC), was established. Women usually served in administrative, personnel or medical roles, in keeping with standard practice in the civilian world at the time. Women moved into operations support, air traffic control, weather observation and radar operations. Nurses entered combat zones during World War II, Korea and Vietnam. Perhaps the most famous female role was that of a flight nurse. One of the most successful of these was Capt. Lillian Kinkela Keil. She flew 250 missions in the European Theater of Operations, crossed the Atlantic 25 times and, during the Korean War, flew 175 more missions. She provided care for 10,000 patients.

Women remained a separate group, WAF (Women in the Air Force), from 1948 until 1976. SSgt. Esther Blake was the first WAF. She was the mother of two bomber pilots and enlisted in the WAC when her boys were shot down over Europe in 1944. Her sons survived the war, and Blake transferred from the Army on July 8, 1948. She served until 1954, then worked for the Veterans Administration. She passed away in 1979.

The first director of WAF was a World War II veteran, Col. Geraldine May. She entered the Army in 1942 at the age of 47 and served in Air Transport Command staff roles. She stayed in the Army after the war and transferred to the Air Force in 1948. She served until 1951.

Jeanne Holm was the first female general. She joined the Army in World War II. Promoted to colonel in 1965, she became the director of WAF. She was promoted to brigadier general in 1971 and retired as a major general in 1975.

Grace Peterson was the first female Chief Master Sergeant. She enlisted in

1942 and became a WAF in 1948. She was among the first NCOs promoted to CMSgt in 1960 and retired in 1963. She and her husband (also a chief) are buried together at Arlington National Cemetery.

In 1967, legislation removed a limit on the number of females in the Air Force. Jane Leslie Holley, from Auburn University, was the first woman commissioned through AFROTC in 1971. That same year, Jeanne Holm became the first woman promoted to brigadier general, and a long-standing policy requiring pregnant women to separate from the military ended. The year 1976 was a great one for women in the Air Force: The WAF ended, and women were simply Air Force members. The Air Force Academy admitted women, and 97 women graduated and were commissioned four years later. In September 1976, the first women pilot candidates entered Undergraduate Pilot Training at Williams AFB, Arizona, and graduated in September 1977. The first five female navigators graduated at Mather AFB, California, the following month.

The first woman to pull missile alert was a Titan II crew member, Lt. Patricia Fornes, at McConnell AFB, Kansas, on September 16, 1978. Fornes would become the first woman to command a combat missile squadron, the 740th Missile Squadron at Minot AFB, in 1993. Her father had commanded the same unit 25 years earlier. In 1985, SAC created all-female Minuteman missile crews. Privacy had never been an issue in launch facilities, so it took a while to incorporate that, but in 1988, mixed male/female crews were introduced in Minuteman and Peacekeeper launch facilities.

Capt. Sandra Scott became the first woman tanker pilot to perform SAC alert duty in 1978. In the 1986 raid on Libya, six Air Force women served as pilots, copilots and boom operators on the KC-135 and KC-10 tankers that refueled FB-111s. On December 14, 1989, women were permitted to serve on C-130 and C-141 airdrop missions, meaning they were aboard tankers and transports in the combat zone during Operation Desert Storm. On April 28, 1993, women were officially no longer excluded from combat roles, as announced by Defense Secretary Les Aspin.

Col. Betty Mullis was the first woman to command a flying squadron (1993) and wing (1996) in the Air Force Reserve. She retired as a major general. The first active-duty woman to command a Flying Wing was Brig. Gen. Teresa M. Peterson, at the 305th Air Mobility Wing at McGuire AFB on March 1, 2002.

Capt. Jacquelyn Parker was the first woman pilot to graduate from the Air Force Test Pilot School at Edwards AFB in 1989. Lt. Jeannie Flynn completed training in an F-15E Eagle to become the first woman fighter pilot in the Air Force on February 10, 1994. She became the first female fighter wing commander when she assumed command of the 4th Fighter Wing in 2012 and rose to major general before retiring in 2023.[29] Then-Lt. Col. Martha

---

[29] Jeannie Flynn Leavitt since her marriage.

Jeannie Flynn Leavitt: fighter pilot, combat leader

McSally became the first female fighter squadron commander in 2004 and was the first female USAF pilot to fly in combat, in the No-Fly Zone over Iraq in 1995.

On November 12, 2024, during an aerial engagement against Iranian drones and missiles, the crew of one F-15E, pilot Maj. Benjamin Coffey and Combat Systems Officer Capt. Lacie Hester, were awarded the Silver Star. She was the first female Airman to receive this prestigious decoration.

Maj. Susan Helms became the first US military woman in space aboard Space Shuttle *Endeavor* on January 13, 1993. Helms would retire as a lieutenant general in 2014. Lt. Col. Eileen M. Collins became the first female Space Shuttle pilot on February 3, 1995. Collins became the first woman commander of a Space Shuttle mission on the *Columbia* on July 23, 1999.

As these pioneers moved up, some women came to top leadership billets:

In 1993, Sheila Widnall became the first female Secretary of the Air Force.

Brig. Gen. Susan Desjardins, a 1980 USAFA graduate, was named the Academy's commandant of cadets on October 7, 2005, the first woman to hold that post.

CMSAF Joanne Bass was the first female Chief Master Sergeant of the Air Force, serving from 2020 to 2024.

Janet Wolfenbarger became the Air Force's first female four-star general in 2012. Since then, four more women have risen to the highest rank: Ellen Pawlikowski, Lori Robinson, Maryanne Miller and Jacqueline Van Ovost. Today, about 21 percent of Air Force members are women, and all opportunities are open to everyone.

General Janet Wolfenbarger, the first woman to wear four stars

MAJOR COMMAND REALIGNMENT

From the birth of the Air Force, there were three major commands in the United States: Strategic Air Command, Tactical Air Command and Air Defense Command. As the Soviet bomber threat lessened (missiles were their primary weapon), Air Defense Command withered, while Military Airlift Command became more important.

On June 1, 1992, SAC and MAC were inactivated. SAC bombers and reconnaissance aircraft went to a new Air Combat Command (ACC), combining with TAC's fighters in a "shooter" command. SAC ICBMs went to ACC at that time; they are now in Global Strike Command. Air Mobility Command (AMC) replaced MAC, taking command of MAC cargo transports and gaining SAC's tankers. Air Force Space Command took charge of the Air Force Satellite Communications System, formerly a SAC asset. The Department of Defense activated the United States Strategic Command, which is responsible for US nuclear forces and long-range delivery systems. Gen. George L. Butler, USAF, the last SAC Commander, became the first commander of United States Strategic Command. On July 1, 1992, Air Force Logistics Command and Air Force Systems Command combined to form Air Force Materiel Command. Exactly one year later, Air Training Command became Air Education and Training Command (AETC).

DESERT STORM

On August 2, 1990, Iraq invaded Kuwait. CENTCOM (Central Command; the regional military headquarters) commander Gen. Norman Schwarzkopf wanted the Air Force "to work on a strategic bombing campaign aimed at Iraq's military..." Col. John Warden was in charge of the "Checkmate" team at the outset of the First Gulf War, the Air Force's war planning unit.

Warden flew the OV-10 Bronco as a Forward Air Controller in Vietnam. Like many officers of his generation, Vietnam instilled in him the necessity for a strategy employing overwhelming force, clear objectives and the integration of political and military aims to attain victory. At the National War College, Warden prepared his first book, *The Air Campaign: Planning for Combat*. Like Billy Mitchell, Warden viewed airpower as the preeminent factor in modern warfare.

Warden's idea for the campaign against Iraq was based on his Five Rings Model: pummeling leadership, key production, infrastructure, population and fielded military forces. He saw an enemy as a "system." Taking out command, control and communications would initiate strategic paralysis. The near-total ineffectiveness of Iraqi forces in the conflict was a powerful demonstration of airpower's capabilities.

JET NOISE IS THE SOUND OF FREEDOM

The United States sent an F-15 squadron from Langley AFB, Virginia, to Saudi Arabia within days of Iraq's attack. Operation Desert Shield then built up forces to deter Iraq and defend Saudi Arabia. President George H. W. Bush activated the Civil Reserve Air Fleet (CRAF) on August 17. CRAF is a contract between the government and airlines to use their aircraft in a national emergency, established in 1952. Desert Storm saw its first deployment. The president then authorized a callup of 200,000 reservists. The Desert Shield buildup continued through the rest of 1990, and President Bush built a coalition of 40 countries to oppose Iraq. Saddam continued his bellicose rhetoric and refused to withdraw from Kuwait.

The war began at 0300 local time on January 17, 1991. More than a thousand airstrikes hit Iraqi targets in the first 12 hours of the campaign, battering the regime's leadership, communications, transportation, air defenses and the electrical power grid. Suppression of enemy air defenses and attrition of Iraqi forces were key. Iraqi ground forces were hit from the outset.

USAF and Army Special Operations forces destroyed enemy radar sites. B-52s hit command and control sites. Other targets were methodically pounded. By January 27, the Coalition had gained air supremacy. Iraqi planes were being buried in the desert or flown to Iran. On January 31, an AC-130 was shot down by a SAM. Fourteen crewmembers were killed.

Coalition pilots destroyed 40 Iraqi aircraft in aerial combat, losing 26 of their own, all but one to AAA or SAMs. The Air Force lost 14 aircraft. Twenty Airmen were killed in action, and six more died due to other causes. One terrible event marred this record of success. On February 13, F-117s hit a communications center that Saddam was also using as a bomb shelter. Perhaps 400 civilians were killed. This brought the air war under great international scrutiny.

The Air Force proved its capabilities across the entire spectrum of air warfare. Ground attack, bombardment and defeating enemy airpower were the most obvious tools, but C-130s moved US Army units to desert locations to set up the maneuver that encircled the Iraqi Republican Guard. RC-135, E-3 AWACS and E-8 JSTARS aircraft monitored Iraqi communications and movements. KC-135 and KC-10 tankers extended the reach of tactical aircraft, enabling a round-the-clock attack.

The ground war began on February 24. The battered Iraqi Army, under heavy bombardment and constant observation, was trapped. More than 2,000 vehicles were destroyed, and about 1,000 Iraqi troops were killed on the "Highway of Death," as more than 3,000 USAF combat sorties struck. The ground war ended after 100 hours of fighting.

The E-8 Joint STARS was still in development during Desert Storm, but was pressed into service. This Joint Surveillance Target Attack Radar System aircraft provided ground battle management and surveillance capability akin to that offered by AWACS for the air battle. JSTARS was retired in 2023, as modern systems may be used on other, smaller aircraft.

The A-10 Thunderbolt II made its bones in Desert Storm. First delivered to Davis-Monthan AFB in 1976, this heavily armored aircraft was built around a powerful 30mm Gatling gun in its nose. It was designed for Close Air Support (CAS) missions. Many in the Air Force were not fans of the slow, hulking, single-mission-focused A-10. But it was proven in supporting ground forces. Built to halt attacking Soviet tank columns, it has been a decisive weapon in air-ground battles since Desert Storm. Though enemy troops fear it, the A-10 is perennially on the Air Force's retirement list. Still, it is unexcelled in its mission.

SOMALIA: To save victims of drought and civil war in Somalia from starvation, the United States launched Operation Provide Relief in August 1992, airlifting food, water, medical supplies and other aid to Somalia and Kenya. On December 4, 1992, President George H. W. Bush initiated Operation Restore Hope, sending 32,000 troops to Somalia to protect relief efforts. The Air Force moved 50,000 people and more than 45,000 tons of cargo. Somali warlords attacked UN forces, and fighting escalated. The Battle of Mogadishu was fought on October 3-4, 1993, as US forces tried to apprehend Somali warlords. Two Black Hawk helicopters were shot down, which forced a desperate defensive stand and a grueling rescue mission. Eighteen American soldiers were killed, and 73 were wounded during nearly 15 hours of intense combat against overwhelming numbers. Pararescueman TSgt Timothy Wilkerson rescued five wounded Army Rangers under fire. He was awarded the Air Force Cross, and 10 other Airmen were decorated for heroism. After this fight, C-5s and C-141s brought 1,300 troops, Abrams tanks and other equipment. Soon after, the Clinton administration released the arrested warlords and withdrew American forces.

BOSNIA: On April 7, 1992, the United States recognized the independence of Bosnia-Herzegovina, Croatia and Slovenia as Yugoslavia broke apart. However, civil war erupted, and by July, US and NATO aircraft began delivering food, medical supplies and other relief to Sarajevo, Bosnia's capital, in Operation Provide Promise, the longest sustained relief operation in Air Force history. In eastern Bosnia, airdrops aided refugees. On March 31, 1993, the United Nations declared a No-Fly Zone over Bosnia, and NATO launched its first combat mission, Operation Deny Flight, to enforce it. Fighting and atrocity continued in this hideous conflict. On February 5, 1994, a mortar attack on Sarajevo's central market killed 68 people and wounded more than 150. In response, the US committed additional aircraft to Deny Flight. Five KC-135s deployed to Istres AB near Marseille, the first stationing of US aircraft in France since the 1960s. The air war heated up on February 28, 1994, when two F-16s shot down four Serb attack aircraft over Bosnia, the first combat in NATO history. Two F-16s struck a Bosnian Serb command post near Gorazde on April 10, 1994, after an attack on United Nations personnel, NATO's first

JET NOISE IS THE SOUND OF FREEDOM

F-16: the standard for tactical airpower

bombing mission. Provide Promise flights took hostile fire—C-141s and C-130s were hit by small arms fire near Sarajevo, and NATO aircraft struck Serb forces in retaliation for attacks. Significant exchanges of fire continued, and a new bombing operation began in May 1995, hitting Serb military facilities in Bosnia. On June 2, Capt. Scott O'Grady, flying an F-16, was shot down by a surface-to-air missile. He evaded for six days before he was rescued. In the summer of 1995, Bosnian Serbs massacred thousands of Bosnians in Srebrenica and deliberately bombed "safe" areas. NATO launched Operation Deliberate Force on August 30, 1995, pummeling more than 300 targets. The Serbs agreed to peace terms by the end of 1995, and talks were held at Wright-Patterson AFB, Ohio. In December, NATO began a peacekeeping operation, Operation Joint Endeavor. American ground and air forces would patrol Bosnia for another decade.

War in the Balkans continued in Kosovo. Operation Allied Force was meant to protect ethnic Albanians from Serb attacks in the Serb province of Kosovo. It was NATO's first combat operation against a sovereign nation. It began on March 24, 1999, and the Clinton administration chose to prosecute this conflict primarily by air. Nearly 1,000 NATO aircraft, more than half from the US, were used. F-15 pilots shot down two Serb MiG-29 fighters on the first day of the conflict. Two days later, one Eagle driver, Capt. Jeff Hwang shot down two MiG-29s. An F-117 stealth fighter was shot down a couple of days later by ground fire; USAF PJs picked up the pilot. C-17s moved supplies to Tirana, Albania, for Kosovar refugees. More than 3,000 tons of food, medicine, tents and other supplies were sent to refugee camps in Albania, Macedonia and Montenegro. The B-2 bomber and the RQ-1 Predator were first used in combat in this conflict. On May 2, a Serbian SAM brought down an F-16 piloted by future Chief of Staff Lt. Col. David Goldfein. He was rescued by a CSAR helicopter. Two days later, an F-16 pilot shot down a MiG-29, the fifth USAF aerial victory in the campaign. On June 10, after a 78-day aerial offensive, Yugoslavia agreed to withdraw Serb forces from Kosovo and to permit peacekeeping forces to protect ethnic Albanians. The complexity of modern air warfare is evident in this operation, as USAF missions included more than 11,000 airlift sorties. Nearly 7,000 tanker missions supported about 9,000 fighter missions, 1,000 Intelligence, Surveillance and Reconnaissance missions and 322 bomber missions. There were almost 1,000 Special Operations sorties and 496 Remotely Piloted Aircraft flights.

As the post-Cold War thaw continued in 1993, an OC-135B Open Skies reconnaissance aircraft was tested for use under a treaty that allows surveillance over treaty nations. In 2004, a Russian TU–154 landed at Travis AFB, California, marking the first Russian Open Skies mission over the United States. This treaty took effect on January 1, 2002.

In November 1994, as part of Project SAPPHIRE, C–5s transported more than 1,300 pounds of highly enriched uranium from the former Soviet Republic of Kazakhstan to the United States to protect it from terrorists, smugglers and unfriendly governments. I think there is an exciting movie plot here.

SADDAM AND IRAQ

US actions in Iraq between Desert Storm and the Global War on Terrorism: Saddam Hussein never really cooperated with UN peace efforts after his ignominious defeat in 1991. Kurds in northern Iraq rose, hoping to win independence, and Saddam reacted fiercely. Gen. Norman Schwarzkopf warned the Iraqis that their aircraft would be destroyed if they flew in this region, and American Airmen downed three Iraqi planes in March 1991. On April 5, the United Nations passed Resolution 688, demanding an end to Iraq's military repression of its people. Operation Provide Comfort established a No-Fly Zone and brought humanitarian relief to the Kurds, enforced by American, British and French aircraft. The US, Britain, France, Germany, Canada and Italy airdropped supplies to Kurds in northern Iraq. Refugee camps were established along the Turkey-Iraq border, with Civil Engineering and Security Police Airmen taking a lead role. As the Kurds were resettled, the mission evolved. Iraqis sought to shoot down Allied aircraft, while the Allies aimed to protect the Kurds and defend themselves, asserting their authority to patrol Iraq north of the 36th parallel. There were numerous incidents of Iraqi challenges to coalition aircraft. These were met with a firm hand: Iraqi air defenses were decimated, and those Iraqis who ventured into the controlled airspace were shot down.

The most tragic day of this operation was April 14, 1994, when F-15s destroyed two US Army Black Hawk helicopters, after a misidentification by AWACS crewmembers and by the F-15 pilots. All 26 people aboard the helicopters were killed.

Provide Comfort ended on December 31, 1996. The mission to contain Iraq continued as Operation Northern Watch (ONW), enforcing the northern No-Fly Zone from January 1, 1997, until the start of Operation Iraqi Freedom in 2003. Some 36,000 sorties were flown. There were frequent periods of air combat as Iraqis fired on Americans, who responded with deadly force. On 28 December 1998, Iraq fired SA-3 surface-to-air missiles against coalition aircraft patrolling the northern No-Fly Zone. In response, Air Force F-15Es, F-16CJs, and Marine EA-6Bs launched anti-radiation missiles and dropped precision-guided munitions on the SA-3 ground-based missile site that had fired on ONW aircraft and destroyed it. From December 1998 to March 1999, US aircraft over northern Iraq came under almost daily fire from Iraqi surface-to-air missile sites and anti-aircraft guns. US aircraft responded by bombing Iraqi air defense sites, using laser-guided bombs as well as AGM-88 HARM missiles and AGM-130 long-range air-to-surface missiles. The first combat use of the AGM-130 was conducted during ONW, when F-15Es deployed a pair to destroy two Iraqi SAM sites. Coalition aircraft flew patrols on an average of 18 days per month, and were usually fired upon. The most common threat was from anti-aircraft guns. Despite Saddam Hussein offering a $14,000 reward for downing a Coalition aircraft, no warplanes were shot down. Low-level conflict continued up until the invasion of Iraq in 2003, although the number of response incidents declined dramatically after 1999.

In southern Iraq, Saddam's violence against the Shiite minority led to another No-Fly Zone in 1992, Operation Southern Watch (OSW). USAF aircraft patrolled Iraqi airspace south of 32 degrees north latitude, protecting Kuwait and Saudi Arabia. On December 27, 1992, Lt. Col. Gary North shot down an Iraqi MiG-25, using an AIM-120 AAMRAM. This was the first aerial victory by an F-16, and the first victory for this advanced missile. Through January 1993, the Iraqis moved SAMs and AAA into the No-Fly Zone, and US planes destroyed many of these. As in the north, there was frequent combat, and US forces destroyed several Iraqi air defense and SAM sites. In October 1994, Iraq again threatened Kuwait, and US forces were sent as a show of force. B-1B bombers flew from South Dakota to a bombing range in Kuwait and back as a show of force. Iraq backed down.

Global flights remained a visible demonstration of American power. In August 1994, two B-52s performed a 47-hour circumnavigation of the earth, dropping 54 bombs on a range near the Iraqi border on the fourth anniversary of Iraq's invasion of Kuwait. Then, on June 2-3, 1995, Operation Global Bat sent two B-1B bombers around the world in a record 36 hours, 13 minutes, 36 seconds, with six aerial refuelings.

On June 25, 1996, terrorists bombed the Khobar Towers housing complex at Dhahran, Saudi Arabia, killing 19 Airmen and wounding 300 Americans. Most USAF personnel in Saudi Arabia were then moved to remote Prince Sultan AB in the Arabian desert. In August 1996, Iraqi troops took Irbil, a Kurdish city in the UN-protected area of northern Iraq. The American response was Operation Desert Strike, sending carrier-based Navy missiles,

B-52: the oldest warrior in the Air Force

planes and B-52s armed with Cruise Missiles against targets in southern Iraq in September 1996. The B-52s struck Baghdad power stations and communications facilities with Conventional Air-Launched Cruise Missiles on a 34-hour, 16,000-statute-mile round trip from Barksdale AFB, Louisiana.

Harbingers of another war were evident: On August 20, 1998, the US launched Cruise Missiles from ships and submarines against terrorist targets in Afghanistan and Sudan in retaliation for the bombings of US embassies in Kenya and Tanzania. Operation Desert Fox kicked off after Iraq refused to permit UN weapons inspectors to do their work. It was a four-day bombing campaign in December 1998, using cruise missiles and air strikes against weapons-production facilities.

These ongoing operations led the Defense Department to find ways to reduce collateral damage. *The New York Times* reported that, "Wary of killing civilians, the Air Force has begun filling 2,000-pound, laser-guided bombs with concrete instead of explosives and dropping them on military targets near populated areas."[30]

In October 2006, the Air Force declared initial operational capability for the GBU–39/B Small Diameter Bomb I, manufactured by Boeing. This bomb, with its 285-pound warhead, had a smaller lethality radius, minimizing the potential for collateral damage.

---

[30] New York Times, October 7, 1999, Section A, Page 3

HAITI: US forces deployed there on September 19, 1994, to restore the country's democratically elected president and stem the flow of refugees to the United States. This effort lasted about 18 months. Air Force support for Operation Uphold Democracy was primarily airlift. In early 2004, Air Mobility Command airlifted Marines into Haiti to restore calm after political violence. A powerful earthquake struck near the capital, Port-au-Prince, in January 2010. About 250,000 people were killed and a million were rendered homeless. AMC's contingency response was immediate. Airlift missions brought humanitarian and aeromedical aid, as well as the 82nd Airborne Division for security and to support recovery operations.

EAST TIMOR: On September 20, 1999, C–130s began airlifting US troops from Australia to East Timor in a United Nations peacekeeping operation. After East Timor declared its independence from Indonesia, a bloody war ensued. Twenty-one C–130s, three KC–135s, two C–17s and a C–141 took part in the operation, which airlifted international peacekeeping troops to East Timor by way of Australia.

JIM PFAFF

CHAPTER 18
WAR ON TERROR

"A second-best Air Force is like a second-best hand in poker—no good at all."
- Hap Arnold

On September 11, 2001, Islamist terrorists hijacked four airliners, crashing two of them into the 110-story twin towers of the World Trade Center in New York, one into the Pentagon, and one was forced down in Pennsylvania amid a heroic counterattack by passengers. The physical damage was substantial. The deaths of 3,000 people and societal changes scar America still. President George W. Bush announced a war on terrorism and initiated an unprecedented homeland-defense effort. For the first time, combat air patrols were flown inside the United States, in fear of further attacks. On September 27, Secretary of Defense Donald Rumsfeld announced that the president had given authority to certain military commanders to order the destruction of hijacked civilian airliners. Many military officers found this curious in light of provisions in the law of armed conflict (zealously enforced in Afghanistan and Iraq) about not targeting civilians.

The Global War on Terrorism (GWOT) was the first in Air Force history in which most decorations for heroism did not go to aircrews. The role of Special Operations was huge. Pararescue, Combat Controllers, Tactical Air Control Parties (TACPs), Combat Weather and Special Reconnaissance (SR) Airmen have been at the forefront of operations. Air Force flying roles were primarily close air support, cargo and personnel movement, and ISR (Intelligence, Surveillance and Reconnaissance).

The initial push after 9/11 was to build homeland defenses and gather intelligence. Fighter planes patrolled the skies over American cities. On October 9, 2001, NATO deployed AWACS aircraft from Europe to the United States to help guard the skies for about seven months after the terrorist attacks. Operation Eagle Assist was the first time NATO forces were deployed to support a member nation under attack.

On October 7, 2001, **OPERATION ENDURING FREEDOM** started with air strikes against terrorist strongholds and Taliban (Islamist regime) targets in Afghanistan. B-2 bombers flew from Whiteman AFB, Missouri, to Afghanistan on 44-hour round-trip missions. That same night, C-17s made their first combat airdrops and delivered food to Afghans cut off by the Taliban. B-1s and B-52s, Cruise Missiles and Navy carrier aircraft struck hard. C-17s inserted Marines and Navy Seabees near Kandahar. Shortly after, Marines were flown into hostile territory on C-17s, landing on dirt strips while using night vision goggles. The campaign was fought in air supremacy from its outset. Afghanistan's few dilapidated aircraft and SAMs were ravaged in the first strikes. Within a few days, F-15E and F-16 aircraft were flying into Afghanistan from Kuwait. Strike aircraft often orbited, waiting to be called into action when there was a TIC call (Troops in Contact). But there were setbacks: in November, an MH-53 crashed on a rescue mission and its crew had to be rescued; a B–1B crashed in the Indian Ocean just off Diego Garcia in December (the crew survived).

The war was global. In January 2002, US and Philippine forces fought Abu Sayyaf, an Islamic terrorist organization operating in the southern Philippines. The Philippine government defeated the terrorists in a long campaign.

In Afghanistan, on February 4, 2002, a remotely piloted MQ-1B Predator aircraft destroyed an enemy target for the first time with a Hellfire missile, killing Al Qaeda leaders. Remotely Piloted Aircraft became a weapon of choice, able to find and kill the enemy without placing Americans in immediate danger.

The enemy's weapon of choice was the IED (Improvised Explosive Device). A homemade bomb designed to kill, destroy or otherwise do harm, IEDs are used by criminals, vandals, terrorists, suicide bombers and insurgents.

Operation Anaconda kicked off on March 1, 2002, as the U.S.-led coalition sought to surround Taliban and Al Qaeda forces. Two Airmen were killed in this action, the first USAF combat deaths of this war. USAF aircraft dropped laser-guided "thermobaric" bombs, explosives with powerful blast and fire effects, especially in confined spaces such as caves and bunkers.

Technical Sergeant John Chapman was a Combat Controller, part of a team deep in enemy territory on March 4, 2002. When a Navy SEAL fell out of a damaged helicopter, Chapman, on the ground, called in an AC-130 gunship to cover the team and search for the fallen SEAL. He brought in another helicopter to extract the team and aircrew, and engaged the enemy, fighting until he was mortally

TSgt John Chapman in Afghanistan

wounded. The SEAL leader credited Chapman with saving the entire rescue team. John Chapman was awarded the Medal of Honor for his heroism.

The **B-1 LANCER** became a powerful tool for helping ground troops. A heavy bomber with variable-sweep wings, it can bring 75,000 pounds of munitions to a fight. In the 1960s, designers sought to develop a Mach 2 bomber with a bomb load comparable to that of the B-52. North American Rockwell won the contract in 1970, but the Carter administration cancelled it. The Reagan administration brought the B-1 back, with some alterations, and in 1985, the first B-1B's were delivered to Dyess AFB, Texas. A total of 100 were procured. At the end of the Cold War, the B-1 was relieved of its nuclear mission and began its conventional bomber career. It saw combat during Operation Desert Fox in 1998 and Kosovo in 1999, then extensive service in the wars in Afghanistan and Iraq.

The B-52 saw action again. In November 1981, eight B–52 bombers displayed their reach in a 15,000-mile, 31-hour sortie with three aerial refuelings from bases in North Dakota, during the Bright Star exercise, bombing a simulated runway in Egypt. This capability was used against Iraq in 1991, at the outset of Desert Storm. B–52G crews from Barksdale AFB, Louisiana, made a 35-hour round trip, launched Cruise Missiles against targets in Iraq and returned home. More than 1,500 B-52 sorties were flown in Desert Storm, pummeling Iraqi ground forces and other targets, delivering 40 percent of all weapons dropped by coalition forces. In the GWOT, the old bomber added to its illustrious history. An Air Force fact sheet said, "In 2001, the B-52 contributed to the success of Operation Enduring Freedom, providing the ability to loiter high above the battlefield and provide close air support through the use of precision-guided munitions."[31]

OPERATION IRAQI FREEDOM aimed to eliminate threats emanating from the Saddam Hussein regime in Iraq, a terrorist ally and weapons of mass destruction suspect. The United States went into this campaign with a long history, through Operations Northern and Southern Watch. These had familiarized American pilots with the area while they destroyed much of Iraq's air defenses. The Iraqi Air Force did not enter the fight in 2003. In fact, some of the Iraqi Air Force was buried in the desert.[32] The Iraqis concentrated their SAMs and AAA in the area between the northern and southern No-Fly Zones, creating a robust air defense for Baghdad and its neighboring cities. However, these would generally be ineffective as they chose not to use targetable radars to aim the weapons. The war commenced on the morning of March 19, 2003, with F-117 strikes attempting to kill Saddam. These did not find the dictator, and a full-scale campaign promising "shock and awe," was launched on March 21. A series of air and cruise missile attacks led the way. Some 700 sorties pounded targets, and a week later, nearly 1,000 paratroopers parachuted from C–17s

---

[31] https://www.af.mil/About-Us/Fact-Sheets/Display/Article/104465/b-52h-stratofortress/

[32] One of these Iraqi planes, a MiG-25, is at the National Museum of the US Air Force.

into northern Iraq, the first combat paratroop drop using C–17s. A vigorous campaign devastated Iraq's air defenses, and by April 8, the US military had achieved air supremacy over the entire country. The ground battle against Iraqi government forces was quick; the counterinsurgency was not. Air Force assets supported the years-long fight with CAS from B-1s and A-10s, with medevac, and with cargo aircraft to facilitate resupply of various bases, which reduced truck usage and the devastating attacks on road convoys.

In March 2011, President Barack Obama authorized a "military action in Libya in support of an international effort to protect Libyan civilians." Enforcing a UN Security Council resolution, international forces struck at Libya in Operation Odyssey Dawn.[33] The operation included establishing and maintaining a No-Fly Zone over the country to protect Libyan civilians. NATO took command of all military operations in Libya under the name Operation Unified Protector. On April 1, in support of this mission, Mobility Air Forces formed the 313th Air Expeditionary Wing in three days and included aircraft from 10 Air Force bases. With KC–135s and KC–10s, they provided aerial refueling to US and coalition aircraft. Four B-1Bs opened the campaign with strikes originating from their base in South Dakota. F-15Es based in Britain conducted most of the interdiction missions, while F-16CJs launched from Germany suppressed Libyan air defenses. On November 3, 2011, NATO's Unified Protector ended.

KC-135 refueling F-15: the greatest tanker meets the greatest fighter

---

[33] People wonder where these names come from. Pentagon planners aim for inspiring and descriptive terms. Desert Shield and Desert Storm were excellent. Others are less so: Desert Fox (German Field Marshal Rommel's nickname) and Odyssey Dawn (is that an exotic dancer's stage name?) were two that the troops lampooned. Names like Strangle or Killer from World War II or Korea are too bellicose for today's world. Future missions will likely be "Freedom this" or "Enduring that" or the occasional "Midnight Hammer."

JET NOISE IS THE SOUND OF FREEDOM

F-22: air superiority fighter

Operation Inherent Resolve began in 2014, aimed at stopping the rapid growth of ISIS hegemony over much of Iraq and Syria. A strong air campaign and support of opposing forces decimated ISIS in 2016-17. B-52s flew another 1,800 combat sorties over Iraq and Syria. The F-22 Raptor flew its first combat sorties in 2014, bombing Islamic State targets. The Lockheed-built aircraft, combining stealth and supersonic speed, first arrived in the Air Force in 2005. In Inherent Resolve, F-22s conducted close air support and proved a valuable ISR platform. In November 2017, F-22s bombed opium production and storage sites in Taliban-controlled regions of Afghanistan. While these campaigns destroyed ISIS in Iraq, the terrorist movement remained like a monster from mythology, with a new head growing whenever an old one was destroyed.

The F-35 saw its first combat on 27 April 2019, in an airstrike on an Islamic State tunnel network in northern Iraq.

August 2021 saw a low point in American military history, as the United States departed Afghanistan, leaving its people to the mercy of the Taliban.

However, Airmen performed heroically, evacuating some 80,000 refugees under conditions ranging from austere to dangerous. The Lockheed-Martin F-35A is the Air Force's latest fifth-generation fighter. It will replace the aging F-16s and A-10s, bringing an enhanced capability to survive in advanced threat environments.

US Air Force F-15E Strike Eagles and F-16 Fighting Falcons shot down dozens of Iranian drones as they were heading toward targets in Israel on April 13, 2024. Over 100 ballistic missiles were airborne at the same time during the attack.

In June 2025, B-2 bombers struck Iranian nuclear weapons sites in Operation Midnight Hammer, with F-35s attacking Iranian defenses. The aggressive nature of the Iranian regime, its avid support for terrorism, and its potential to create massive regional chaos and conflict made reducing Iranian nuclear capabilities a key part of any Middle East peace strategy.

JIM PFAFF

CHAPTER 19
COME FLY WITH ME

*"If you're going to fly airplanes,
the best place to be is the Air Force."*
- Astronaut Michael P. Anderson

Let's go flying! We have seats on a KC-135 tanker; that's the airplane I know best, and tanker crewmembers are my "Bubbas" (people who fly the same aircraft) so I don't expect any "buffoonery." It means clowning around, but fliers use it as a catch-all for the dumb stuff people do that lands them in hot water.

Welcome to the squadron. You'll hear some unfamiliar terms in conversation here. A term appearing frequently in this dynamic environment is "line of sight scheduling." One of the most dreaded sights is the operations officer prowling the halls, looking for someone who can take on some undesirable task or deploy unexpectedly and very soon. People avoid eye contact and maybe slip out the door, but someone's about to have a different day than they planned.

KC-135 air refueling C-5: profile view

KC-135 air refueling C-5: boom operator view

Another term we hear in a flying squadron is "lost the bubble." It means someone got confused or misunderstood what was happening. The term comes from navigation: Sextants had a liquid-filled chamber that the navigator sighted through. A bubble was kept centered, ensuring the sighting was measured against a level plane. Losing the bubble meant you had no reference to level, and your sighting would be inaccurate.

If you're in a big airplane, in formation, someone has to be "Tail-end Charlie." That is just life; someone's always bringing up the rear. But if you get labeled "Tail-end Charlie" on the ground, you're consistently behind your peers.

Fighter planes often fly low-level, down "in the weeds." The term also means accessing deep layers of information, often more than was needed: "He started talking about his flight last week and next thing you know, he's down in the weeds."

Fighters and bombers may drop "iron bombs"—old-fashioned "dumb" bombs, targeting the enemy with "iron on the target." There are JDAMs, or Joint Direct Attack Munitions—smart bombs, which are essentially dumb bombs with a GPS guidance package added.

Fighter pilots may practice "Dissimilar Air Combat Training" (DACT), training between two different types of aircraft. Ideally, this provides missions in which an "adversary aircraft" uses the tactics of potential enemies.

Before a combat mission, we may discuss an enemy IADS (Integrated Air

Defense System). It combines air surveillance, battle management, and weapons to minimize or eliminate threats. This may include Jammers (electronic systems that jam, interfere with, or block enemy sensors). Many Electronic Counter Measures (ECM) are available but EW (Electronic Warfare) is a continually evolving science, using electromagnetic energy to trick, evade, spoof or attack enemy forces. If you can impede the enemy's ability to communicate, gather intelligence and coordinate their operations, you have a valuable edge. Some aircraft have dedicated EW experts, called Electronic Warfare Officers (EWO), but informally known as Ravens, Crows or E-Dubs.

Fighter pilots may "jink," maneuvering sharply to avoid a threat, or the air combat becomes a "knife fight in a phone booth," a close-in aerial dogfight. On the ground, jink means to dodge something unwanted: "The colonel was looking for someone for that midnight flight but I was able to jink and he didn't ask me."

We won't need a "Piddle Pack," because there is a latrine on board the tanker. But the Piddle Pack enables fighter pilots to empty their bladders in the confines of their jets. It's difficult to use if you're wearing a "Poopy Suit," the anti-exposure suit worn over the flight suit on long overwater flights. Why is it called a Poopy Suit? It stinks after several hours of wear.

If you're running late, a pilot may say, "kick the tires and light the fires," meaning to get airborne quickly, but on the ground, it means "hurry up!" Someone might tell you to "put it in afterburner!" Afterburners pour fuel into a jet engine's exhaust stream, using the remaining oxygen to burn it, heating and expanding the exhaust gases to increase thrust.

There was some bad weather on the route earlier, but we're going to "press on," and continue our tasking. Before we fly, we'll try to contact the receiver aircraft, our "playmate" for today, the aircraft we work with during our "playtime"—the time an aircraft can remain on station. (Not many people call their jobs "playtime") For fighters, this is especially important; they can't carry a lot of gas. Fighters need to script their scenarios so they don't waste training or create a safety hazard.

We'll stop by Life Support and grab a quick-don oxygen mask. Helmets and parachutes used to be a requirement. Today, most aircrew rely on the mask. With its elastic straps it can be on the face in seconds. On the tanker; the weight of the parachutes has been deemed unnecessary, since KC-135 bailouts have been almost as rare as unicorns in a zoo. Still, jets with ejection seats—fighters and bombers—require helmets and parachutes. That requires additional training: People have accidentally ejected themselves, and that's deadly.

If you fly a pointy-nosed, highly maneuverable, go-fast jet, you wear a G-Suit, occasionally called "speed jeans." It's a chaps-like nylon garment worn around the legs and abdomen, with air bladders linked to a G-sensitive valve, causing the bladders to inflate and apply pressure to keep blood from draining away from the brain during high-G maneuvers.

But the everyday flight attire is the flight suit, known as "the bag." The flier steps into these coveralls, pulls up the zipper and is good to go. It is the most comfortable uniform, like going to work in pajamas. Flight-suited Airmen are a special breed—sleeves rolled up, wearing aviator sunglasses. Many dreamed of flying from childhood, so they are often cheerful, even exuberant. Their accoutrements are leather jackets and a flight cap sticking out of an unzipped leg pocket. Sometimes they wear colorful scarves around their necks, a throwback to early aviators who wore scarves to minimize chafing from the constant "head on a swivel" of combat flight. Today, scarves are usually worn because the commander said so. Fliers, particularly fighter pilots, are the rock stars of the Air Force... and they know it. They are occasionally full of themselves and maybe a little obnoxious. "Nonners" sometimes hate fliers, and fliers don't care. Some who don't fly call fliers "zipper-suited sun gods." The movie *Top Gun,* about Naval aviators, captured the flight suit attitude.

Every Air Force mission begins with extensive preparation. Training events are a focus, but safety is paramount. Airplanes are expensive and people are irreplaceable. Operational Risk Management (ORM) is discussed. ORM evaluates options, risks and benefits to determine the best course of action. That's good advice for anyone. ORM principles include:

1. Accept no unnecessary risk without a commensurate return.

2. Make risk decisions at the appropriate level. Those accountable for the success or failure of the mission must be included in the decision process.

3. Accept risk when benefits outweigh the costs. Potential benefits should be compared to potential costs. Some high-risk actions bring significant returns.

4. Integrate ORM into operations and planning at all levels.

Training missions are often scripted within the unit: going to a range to perform combat tactics, or to a drop zone, an assault strip, or air refueling. Operational missions are planned for the units in the FRAG, the "fragmentary" orders derived from an overall Air Tasking Order (ATO). Aviators tend to refer to their role in the ATO as being "fragged," and a mission flown as planned "went as fragged."

Perhaps the training profile calls for an Instructor Pilot (IP) to fly with a crewmember, or someone is due for a flight examination with an EP (Evaluator Pilot) from Stan/Eval (Standardization and Evaluation). Most Airmen fly each day as if it were a "checkride," but with someone evaluating you, it's possible to "bust" (fail) the ride. Sometimes a failure is called a "hook," from the shape of the letter U, as in unsatisfactory. It's a blemish on your record to bust a checkride. But it happens. Fortunately, a doghouse is only one deep, so the damage may be limited if someone else gets their name in lights.

WELCOME TO THE CREW

Our crew is a good one. The aircraft commander is known for "golden hands" (a pilot who flies really well), and the copilot is a "good stick" (he flies well). They have the "gouge" (the latest information), after checking the Read File and the FCIF (Flight Crew Information File). These Files have useful information—a navigation aid is inoperative, or there may be a work crew on the runway, so ask that the runway is clear before landing, etc. Fortunately, we don't have a "ham fist" flying today, a pilot who is rough on the controls. The Boom Operator is the only enlisted crewmember on the tanker. A Boom Operator plays many roles, such as loading aircraft cargo or managing passengers. But mostly, they "boom." The boom is a 28-foot-long tube located at the rear of a tanker. It has an extendable nozzle inserted into the refueling receptacle of a receiver aircraft, enabling jet fuel to flow at nearly 1,000 gallons a minute. In the KC-135, the boomer lies on a pallet in a pod with windows at the rear of the aircraft, on the lower side of the tanker's body. In the KC-46, the "Inflight Refueling Specialist" (the formal title) sits at a console in the cockpit, using cameras to project the image of the receiver aircraft onto cockpit displays.

In any case, we will go over Emergency Procedures (EP). Every aircraft has EPs that a pilot or crew must follow in case of system failure, fire or other emergencies. In-depth systems knowledge—engines, hydraulics, electrics, avionics and other components—is vital to the safety of any aircraft operation. Some EPs must be memorized verbatim. These are printed in boldface in the flight manuals, and crewmembers are frequently tested on them. Today, we will plan to fly under Instrument Flight Rules (IFR), in case visibility, visual contact with the ground or weather patterns impact flight. Most light aircraft you see flying are using visual flight rules (VFR). We will discuss MARSA (Military Accepts Responsibility for Separation of Aircraft) on the Air Refueling (AR) Track, since we will actually be in contact with another plane in flight. There are AR Tracks all over the United States, and sometimes the refuelings are visible from the ground. We may discuss Controlled Flight Into Terrain (CFIT). Flying an airplane into the ground is fatal. Gravity and the Earth always win. We'll discuss what to do in the event of an IFE (In-Flight Emergency). Someone might mention the "pucker factor," that involuntary tightening of the body in a moment of fright: "The pucker factor went way up when we saw the engine on fire." An IFE is uncommon, but any inbound aircraft with an IFE means there will be a parade of fire trucks, ambulances and maintenance vehicles toward the flightline to assist if needed.

The KC-135 doesn't have ejection seats. If it did, the pilot might have referred to the possibility that you might "punch out," or eject from the aircraft. In this jet, it's a figure of speech meaning to get out quickly. It's also used when someone is getting out of the Air Force: "Are you reenlisting?" "Nope, I'm punching out." So, if a pilot mentions punching out, he may have airline applications in his bag.

You'll hear "Roger" frequently in conversation among fliers and with other aircraft and air traffic control. It means the listener has received the message, understands it, and will follow it. You may hear "wilco," which is short for "will comply," or someone may say "copy," acknowledging what was just said ("I heard and understood the message"). It becomes part of the day-to-day vernacular of a flier: "I'll meet you at the restaurant at 1200 (pronounced twelve-hundred)." "Roger."

We will calculate a "bingo fuel," the minimum fuel for a safe return to base. In combat, aircraft may fly past bingo fuel, but doing so is risky. It may mean an emergency air refueling or "diverting" to another base.

If we were flying operationally—not a training sortie—we might be part of a "package," a team of aircraft which may be co-located or geographically separated units. It could be a "coronet," refueling deploying fighters. It could be a "gorilla," a large force package. While it's common to launch from several bases and position as required, sometimes an entire force launches from one base. It's impressive to see dozens of aircraft launch one after the other: AWACS, counter-air fighters, jammers, strike fighters (carrying bombs), and refuelers.

The flier "steps" to the jet. "Step time" is when we go to the bus or van. Missing step time can mean you're walking to your plane. A multi-place aircraft may see lots of people boarding the jet. The people are "crewdogs," the team that flies with little of the bravado of the single-seat pilot. Crewdogs include pilots, copilots, navigators, flight engineers, loadmasters, flying crew chiefs, gunners, electronic warfare officers, boom operators, airborne maintenance and other specialists, all ensuring the heavies, bombers, airlifters, reconnaissance and tankers accomplish the mission. They may carry bags and publications with them. Some of these are mission binders detailing today's mission plan, including the flight plan, communications data, fuel usage plan, and other pertinent information. There will be "the pubs," the flight information publications with air route charts, airfield information, diagrams of runways and taxiways, and "let-down books, the charts depicting the instrument approaches to runways.

One publication they carry is the "Dash-1," a massive book that serves as the aircraft's "owner's manual." It is a technical order covering aircraft systems and their operation. It explains normal and emergency procedures, crew duties, operating limits, etc. In printed form, it is an encyclopedia-like set of books weighing more than 20 pounds. They are kept in loose-leaf binders to facilitate changes. Even a veteran aircraft like the KC-135 sees systems updates and improved procedures. Some changes are driven by close calls, mishaps, and sometimes loss of life. Old-timers said, "The Dash-1 is written in blood." Aircrew checklists, performance data charts, cargo load data, air refueling data and emergency information are drawn from this enormous resource. Aircrews today carry a digital version of this. Its name, Dash-1, comes from its listing as Technical Order (T.O.) 1B-52H-1 (for the B-52H) or 1T-37B-1 (for the T-37B), etc.

Going onto the flightline, we pass through an Entry Control Point (ECP). Identification is mandatory, and a specific badge may be required. The ECP is a well-marked point where vehicle tires are checked for FOD (Foreign Object Debris). Small objects can be sucked into jet engines, causing serious damage. An ECP is serious business: It is guarded by armed Defenders, and some areas are marked as "Use of Deadly Force Authorized." They aren't kidding. The restricted area is bordered by a red line painted on the ground. When someone crosses that line, this is called "breaking red." The offender will find himself face down on the tarmac until someone vouches for him.

Preflight may take 30 minutes to an hour, depending on the aircraft. The mission itself will also vary. Training missions in fighters are typically one to two hours. Tankers and transports will fly two to five hours, depending on the distance to their training area. Bombers, reconnaissance and electronic warfare aircraft may fly longer missions to exercise all their mission sets. It's not unusual for a B-52 crew to take off before dawn and land after sundown. After landing, there is a debrief with maintenance, during which any problems with the plane or its systems are discussed. After an operational mission, there will be a discussion with intelligence, reviewing what happened and where.

While we fly, you may see some amazing technology, but if you ask about it, you might be told it's FM (freaking magic) or even PFM (pure freaking magic). That term usually refers to something that works incredibly well, but it's also applied to a technical problem that somehow resolved itself.

As we preflight radios, we may be told to "go clear" or "go secure," which means using either unencrypted or encrypted voice communications. Secure communication is vital. We may be requested to authenticate information, which tests our ability to use codes to verify our identity. We will also check our IFF—Identification Friend or Foe (IFF) transponder. This identifies aircraft and determines their bearing and range from a known point. It was developed during World War II and air traffic controllers use these "squawks" to identify air traffic.

During preflight, we will program the GPS (Global Positioning System) with waypoints to track our progress and maintain "Situational Awareness." SA is knowing what your airplane is doing relative to its capabilities, where your adversary is, where you are, the status of any threats, and other variables. On modern aircraft, a tool for maintaining SA is the HUD (Head-Up Display), a transparent screen in front of the windscreen (what we would call a windshield on a car) on which flight instrument and weapons system data are projected, keeping the pilot's eyes outside rather than inside the aircraft.

Without SA, it's easy to get "behind the power curve." This refers to a dangerous situation in an aircraft where the required thrust to maintain altitude and airspeed exceeds the available thrust. The term is also used when someone is struggling to keep up with their work. You don't want to be behind the power curve in an airplane or in your life. The term "SA" is often used in the Air Force, and people may send a message to a superior headed "FYSA," or "For Your Situational Awareness," as an update on an issue or a project.

Soon comes the word to the crew chief to "pull chocks," it's time to taxi. Aircraft tires are chocked to keep them from moving. It's surprising how a little wind can move a parked airplane. Airmen may say "I'm pulling chocks" when they mean "Let's go," or "I'm leaving now."

Every aircraft that flies has a call sign: Pearl 51, Reach 907, Air Force One, etc. Some are static call signs used for local training, while headquarters assign others for various missions. Many fliers have personal call signs. Some are plays on a name: Smith may be "Smitty," but Wright will be "Orville" or "Wilbur." Some are based on personal traits or features, or perhaps a resemblance to a well-known character, and often not flattering: "Baldy" (obvious), "Beak" (a large nose), or "Rodman" (newly dyed hair). A guy who wiped out on a bicycle may become "Skid," while one whose heavy foot caught the Highway Patrol's attention might be "Speed." Not everyone gets to be "Maverick" or "Iceman." Your squadron decides that. If you're lucky, your call sign doesn't come from guys who've been drinking!

Takeoffs and landings are always exciting, and newcomers are often astonished by the number of tasks a flight crew performs and monitors. A visitor on a flight may feel like "a pig looking at a wristwatch" or "a dog watching TV"— befuddled, seeing what is happening but unable to act. It's never good when a crewmember does that: "We're lost, and instead of looking at the map, Snuffy just sits there like a dog watching TV."

A tanker and its receiver meet at a pre-determined time and altitude, on an Air Refueling Track. It has a designated entry and exit, with a join up location, the Air Refueling Control Point. We cruise to the AR track, orbit at the Air Refueling Control Point (ARCP), and then our receiver comes down the track, and we turn to roll out ahead of the receiver. Today it's a C-5, and there are several pilots aboard that plane, all wanting to renew or attain "currency" in air refueling. It is a required skill, so receiver pilots must stay in practice because they will need it when crossing an ocean or going into combat. The C-5 is a giant, and its bow wave is like that of a large ship approaching a harbor. It pushes the airliner-sized KC-135 around, and the tanker crew will work hard to stay in position and in contact. The receiver will close slowly, coming in from a thousand feet lower, making visual contact, and slowly climbing to the tanker. When they are close enough, the boom operator makes contact and fuel flows. On a training mission, it may be just a splash to wet the seals in the equipment, or just a "top off," to give a fighter more playtime.

In an operational AR, the fuel flows at a rate of 6,000 pounds per minute. C-5s and B-52s can hold more than 300,000 pounds of fuel. The KC-135 could fill your bathtub in about three seconds or a swimming pool in about 15 minutes. But it'll take longer to fill one of those large aircraft. The skill required to keep these aircraft in contact is impressive, and when the AR goes for 20 or 30 minutes, it is exhausting for all concerned.

When we land, let's grab a cold drink and some jalapeño popcorn. Snack bar officer ("Snacko") is a vital, mission-critical position usually held by new lieutenants in a flying squadron. There is often a popcorn machine, snacks, a refrigerator full of drinks, etc. Doing well as Snacko makes a good impression and leads to better opportunities. Failing will cause ridicule and ostracism.

There was a time when fliers would gather at the bar (within the squadron) or the club on base to discuss the mission informally. Alcohol is less common today, but "hangar flying" still exists, as fliers tell stories and share experiences. It is both bonding and teaching, passing on lessons learned.

THE UNSUNG HEROES OF MAINTENANCE

The aircrew debriefed with maintenance after we landed. Aircraft maintenance troops are meticulous, professional and quietly dedicated. Everyone remembers Charles Lindbergh, but his mechanic's name is forgotten. Maintainers don't seek glory; they make the Air Force fly. Whether they work on aircraft, missiles or munitions, they do challenging work under austere conditions, desert heat, jungle humidity, Arctic cold, wind, rain and snow. They can make machines and their components come to life.

A maintainer briefs the flight crew before every flight. The crew chief will go over the Form 781—commonly called "the 781"—with the aircraft commander and crew. This loose-leaf binder contains current maintenance write-ups, maintenance history, fuel load and other information on the airplane's condition. It is to an aircraft what a medical chart is to a patient. After it's reviewed, it is signed off by a maintenance official and the pilot in command. It's determined whether the plane is FMC (Fully Mission Capable), able to perform all its missions; PMC (Partially Mission Capable), able to perform some missions—perhaps a tanker that can't refuel but could carry cargo; or NMC (Non-Mission Capable), unable to do anything. The crew chief will tell you if anything is "INOP"—Inoperative/Inoperable—and might say the equipment is TANGO UNIFORM or TU. This means it's broken. TU is short for "tits up"—a crude way of saying "flat on its back." A crew chief may be young and junior in rank, but is respected by all. In World War II, a crew chief was typically the lead mechanic on a maintenance team. Now, it refers to an individual responsible for an aircraft's day-to-day maintenance and upkeep. The Dedicated Crew Chief (DCC) for a given aircraft may have several assistants. Crew chiefs may be thought of as "family doctors" who take care of most of the aircraft's ills. At the same time, they will need to call in a specialist for detailed work in specific areas, such as avionics, hydraulics, electrical, environmental, engine, fabrication, structural, metals, etc. Like doctors, the specialists are extensively trained in their particular area. These are among the most challenging jobs in the Air Force. Fuel systems maintainers were featured on the television program "Dirty Jobs." Crawling around in a confined space, wearing a respirator and working with toxic substances isn't for everybody.

An uncomfortable moment comes if the maintainer says, "I'm not the regular crew chief." These words suggest the maintainer was put on this task without much preparation. You may hear this phrase elsewhere on base if someone is responding to an inquiry in an area that's not their usual role. It doesn't give you a "warm fuzzy" security blanket feeling of confidence.

There are various pieces of equipment around the aircraft: Aerospace Ground Equipment (AGE) includes power generators, air conditioner units, heater carts, portable lights, tow vehicles and airstairs; these support the plane before it is on its own power.

Depot Maintenance is in-depth maintenance for which aircraft are sent off-station about every five years or so, for the "inspection and correction of defects that require skills, equipment or facilities not normally possessed by operating locations." It is work—disassembly, inspection, repair, rebuilding, painting and flight testing—beyond the scope of field units. Because it is planned, it is often referred to as PDM (Programmed Depot Maintenance). This is in addition to "phase maintenance," where an aircraft is hangered locally for a comprehensive check and overhaul as required, typically about once a year. People marvel that the Air Force flies half-century-old aircraft. If you took care of your car the way Airmen take care of aircraft, your car would last fifty years, too.

You may hear the term "cannibalization," but don't worry. The Air Force is here to serve man, but most Airmen don't bite. Cannibalizing is the use of parts from one piece of equipment, typically a broken aircraft, to bring another into working order. "Hangar queens" are airplanes that are chronically in hangars for repair. They may become "cann birds," donating parts to other, less-broken planes.

A Centralized Intermediate Repair Facility (CIRF) is an intermediate-level maintenance facility that supports deployed forces. This maximizes resources by keeping assets close to the combat theater while reducing the number of locations needing parts and critical specialists. A Centralized Repair Facility (CRF) is a small, specialized repair and maintenance tasking, often conducted at Guard or Reserve units. A unit may take on the care of a single component—engines for the C-130H or B-52 hydraulics—that provides a cost-saving and efficient logistical solution for the Air Force.

The smell of jet fuel is in the air. POL (Petroleum, Oil, & Lubricants) is a traditional handle for the Fuels Management Flight, which stocks, manages, and issues fuel. Their rally cry is "Who the hell? POL." Without them, the engines don't burn, the wheels don't turn and the birds don't fly. Jet fuels known as JP-4, JP-5 or JP-8 are kerosene-based fuels, similar to civilian Jet A/A-1 fuels but with corrosion inhibitors and anti-icing additives. JP-8 is a less hazardous successor to JP-4, a highly volatile 1950s-era fuel made from kerosene and gasoline. The smell of jet fuel often makes former aviators nostalgic for flight. It can make a flier's daughter think of her dad.

If you're on a plane with weapons—bombs, missiles, guns—you will work with Ammo troops, whose élan sets them apart. They assemble, move, load, arm/disarm and store the bombs and missiles that deliver airpower to the enemy. It's a hazardous job demanding precision and teamwork. They are high-spirited about their potentially dangerous jobs. Without them, the USAF doesn't put iron on target. Their battle cry is IYAAYAS ("If you ain't Ammo, you ain't sh!t!"). Maintainers, pushing hard to get jets in the air and move on to their next task, will occasionally mutter, "IYAAYWOT," which means, "If you ain't Ammo, you're waitin' on them."

You may see some people walking shoulder to shoulder down the runway or a taxiway. They're performing a FOD (Foreign Object Debris) Walk. Hundreds of Airmen walk along taxiways, parking ramps and down the runway, picking up whatever FOD they find. On a pleasant day, it's fun. But somehow, it usually rains. FOD causes aircraft damage. Simple things like a rock, a dropped tool or aircraft part, even a cap or glove, can cause costly damage to an airplane.

There's a great book about aviation in Alaska. Its title: *Flying Beats Work.*[34]

It sure does!

---

[34] *Flying Beats Work: The Story of Reeve Aleutian Airways* by Stan Cohen

JIM PFAFF

CHAPTER 20
TOWARD THE FUTURE

"If we maintain our faith in God, love of freedom, and superior global air power, the future looks good."
- Curtis LeMay

The **B-21 RAIDER** is the follow-on to the B-2 bomber. The first test models are flying, and it is expected to be operational in 2027.

The **KC-46 PEGASUS** continues to have growing pains, but is still forecast to take over the air refueling mission in the early 2030s.

The venerable E-3A Sentry was expected to be replaced by the Boeing E-7 Wedgetail by 2030, but a decision to go with the Navy's E-2D Hawkeye made the news in September 2025.

The first USAF **CV-22 OSPREY** arrived at Edwards AFB, California in October 2000. The Osprey could take off like a helicopter and fly like an airplane. The Air Force's first operational Osprey came in 2006. While it has received bad publicity for a sketchy reliability record in other armed services, the Air Force has found the Osprey a valuable Special Operations aircraft.

Remotely Piloted Aircraft (RPA): In September 1996, the 11th Reconnaissance Squadron became the first Air Force unit to operate the General Atomics RQ-1 **PREDATOR**, an unmanned aerial surveillance and reconnaissance aircraft. As reconnaissance platforms, they served in Bosnia and Kosovo. Weapons came to the Predator in February 2001, when a Predator destroyed a ground target with a Hellfire missile on a Nevada range. They were used in combat in Afghanistan later that year. With a ceiling of 25,000 feet and a 400 nautical mile range, it could stay airborne for 24 hours at 73 knots. The MQ-9 **REAPER** came into service in 2007. It can operate up to 50,000 feet and fly up to 300 knots. With substantially greater precision-guided munitions, camera and radar capabilities, it is an enduring weapons and ISR platform.

In November 2009, AFSOC stopped calling these craft unmanned aerial systems (UAS) and began using the term Remotely Piloted Aircraft (RPA).

Predator Remotely Piloted Aircraft

The Northrop Grumman RQ-4 Global Hawk first flew on February 28, 1998. It was designed for high-altitude, long-range, long-endurance reconnaissance missions. With a wingspan of 116 feet, it could fly at an altitude of up to 65,000 feet and photograph an area the size of Korea in 24 hours. In April 2001, a Global Hawk completed the first nonstop crossing of the Pacific Ocean by a UAV, from Edwards AFB to Australia (7,500 miles) in about 23 hours. They have been used in the GWOT, in Libya and over the earthquake-damaged Fukushima nuclear plant in Japan.

On June 20, 2005, the Air Force redesignated Indian Springs Air Force Auxiliary Field, Nevada, as Creech AFB, the first base primarily dedicated to UAV operations. On May 1, 2007, the Air Force activated the 432d Wing at Creech, the first unmanned aircraft systems wing.

In 2011, Defense Secretary Robert Gates announced that the Air Force was now training more pilots for Remotely Piloted Aircraft than for any other weapons system. In 2025, there are more than 300 of these aircraft in the USAF inventory.

The X-45A unmanned combat air vehicle flew for the first time at Edwards AFB, California, in May 2002. It was the first unmanned aircraft designed for combat operations in a hostile environment. By 2025, the F-47 was in testing, an optionally-manned aircraft that will have unmanned "wingmen," directed by the F-47 systems for high-risk missions.

SUMMARY

The future comes whether we are ready or not. Some say the last fighter pilots have already been born. RPA technology grows better every year: more capable, and more deadly. Drone technology, providing cheap swarms of lethal weapons, will change warfare as much as the airplane did. And yet, my grandchildren may fly B-52s and KC-135s. And so, the Air Force mission continues. Every day, Airmen stand ready in the defense of the United States. Around the world, USAF aircraft "show the flag." As the future unfolds, the men and women of the United States Air Force will carry on the mission left to us by those brave Americans who came before. As James Stewart said in *Winning Your Wings*, "America, your future is in the sky!"

JIM PFAFF

ACKNOWLEDGMENTS

So many people have inspired, encouraged, and helped me along the way. My appreciation to all of you for what you have meant in my life is beyond words. Thank you all so much.

Col. Jim Heathcote, Steve Apple, Gary Picklesimer, Tim Scott, Andy Knapp, Rick McQuiston, Jim Barber, Chuck Schmitz, Dave Katai, Joe Contino, Ron Kaufman, Steve Fout, Mike Byers, Tom Childers, Bennie Branham, Gary Ball, Rob Tyler, Gavin Halsey, Jayden Richards, Colin Chafin, James Hobson, Sasha Bouaroy, Kristin Tully, Wyatt Phillips.

And most of all my family: Denise Pfaff, Johanna Pfaff, Karin Householder, Austin Householder, Blake Householder, Brady Householder, Max Ritualo, Dave Ritualo, Vikki Wooley, Rhonda Tyree-Ryan, Bruce Pfaff. And those no longer with us, but always in our hearts: Jim Pfaff Sr., Lou Ann Pfaff, David Pfaff, TSgt. Bill Franklin, Arthur Tyree, Allie Tyree.

ABOUT THE AUTHOR:

Jim Pfaff is a retired US Air Force colonel. As a navigator with 3,500 flight hours, including combat and combat support time, he flew the KC-135 and RC-135. He was also an aircraft maintenance officer and a squadron and group commander. Jim is a veteran of operations in Panama, Bosnia, the No-Fly Zones over Iraq and Operation Enduring Freedom.

JIM PFAFF

PHOTO CREDITS

1. Author's Collection
   A toy B-58 that started a journey

2. https://commons.wikimedia.org/wiki/File:Signal_Corps_Dirigible_No_1_afmil-01.jpg
   http://www.nationalmuseum.af.mil/factsheets/factsheet.asp?id=665
   First Airship in US military service

3. https://commons.wikimedia.org/wiki/File:Benjamin_Delahauf_Foulois_in_flying_helmet.jpg
   Image of Benjamin Delahauf Foulois from the public domain book A Concise History of the U.S. Air Force by Stephen L. McFarland.
   Benjamin Foulois

4. https://commons.wikimedia.org/wiki/File:Billy_Mitchell.jpg
   http://www.af.mil/shared/media/photodb/photos/020903-o-9999b-081.jpg
   Billy Mitchell

5. https://commons.wikimedia.org/wiki/File:RickenbackerUSAF.jpg
   http://www.af.mil/information/heritage/person.asp?dec=&pid=123006466
   Eddie Rickenbacker

6. https://commons.wikimedia.org/wiki/File:RAF_Chelveston_-_305th_Bombardment_Group_-_B-17_Flying_Fortress_Weary_Bones.jpg
   http://www.americanairmuseum.com/media/338
   This work is in the public domain in the United States because it is a work prepared by an officer or employee of the United States Government as part of that person's official duties under the terms of Title 17, Chapter 1, Section 105 of the US Code.
   B-17 with crew

7. File:Boeing P-26.jpg - Wikimedia Commons
   This image or file is a work of a U.S. Air Force Airman or employee, taken or made as part of that person's official duties. As a work of the U.S. federal government, the image or file is in the public domain in the United States.
   P-26

8. https://commons.wikimedia.org/wiki/File:B29.jpg
   United States Air Force picture, Public domain, via Wikimedia Commons
   B-29

JIM PFAFF

9. https://commons.wikimedia.org/wiki/File:340th_EARS_fuels_Bomber_Task_Force_mission_201210-F-ER377-9634.jpg
Staff Sgt. Trevor McBride, Public domain, via Wikimedia Commons
B-52

10. https://commons.wikimedia.org/wiki/File:F-15C_Eagle_from_the_67th_Fighter_Squadron_at_Kadena_AB_is_refueled_by_a_KC-135R_Stratotanker_from_the_909th_Air_Refueling_Squadron_.jpg
U.S. Air Force photo/Tech. Sgt. Angelique Perez, Public domain, via Wikimedia Commons
KC-135 refueling F-15

11. https://commons.wikimedia.org/wiki/File:Lockheed_C-130_Hercules.jpg
U.S. Air Force photo by Tech. Sgt. Howard Blair, Public domain, via Wikimedia Commons
C-130

12. https://commons.wikimedia.org/wiki/File:P-51D-Mustang-12AF-15AF-325FG319FS-98-Italy-1945.jpg
USAAF, Public domain, via Wikimedia Commons
P-51

13. https://commons.wikimedia.org/wiki/File:B-17Gs_381st_BG_en_route_to_target_c1944.jpg
USAAF, Public domain, via Wikimedia Commons
B-17s in flight

14. https://commons.wikimedia.org/wiki/File:F-86_in_korea.jpg
USAF, Public domain, via Wikimedia Commons
F-86

15. https://commons.wikimedia.org/wiki/File:McDonnell_F-4C_Phantom_IIs_of_the_558th_TFS_in_flight_over_Vietnam,_in_December_1968.jpg
USAF, Public domain, via Wikimedia Commons
F-4s in flight

16. https://commons.wikimedia.org/wiki/File:USAF_F-16_without_background.png
U.S. Air Force photo by Master Sgt. Don Taggart, Public domain, via Wikimedia Commons
F-16

17. https://commons.wikimedia.org/wiki/File:Raptor_F-22_27th.jpg
Technical Sergeant Ben Bloker (USAF), Public domain, via Wikimedia Commons

JET NOISE IS THE SOUND OF FREEDOM

F-22

18. https://media.defense.gov/2005/Mar/10/2000584431/-1/-1/0/050310-F-JZ508-658.JPG
    Public domain This image or file is a work of a U.S. Air Force Airman or employee, taken or made as part of that person's official duties. As a work of the U.S. federal government, the image or file is in the public domain in the United States.
    Curtis LeMay

19. https://commons.wikimedia.org/wiki/File:Air_Force_Chief_of_Staff_General_Curtis_LeMay_with_Secretary_of_Defense_Robert_McNamara_at_The_Pentagon_in_1963.jpg
    Pentagon Archive, CC BY-SA 4.0 <https://creativecommons.org/licenses/by-sa/4.0>, via Wikimedia Commons
    LeMay & McNamara

20. https://commons.wikimedia.org/wiki/File:Robin_Olds_during_vietnam_war.jpg
    Official USAF Photograph, Public domain, via Wikimedia Commons
    Robin Olds

21. https://commons.wikimedia.org/wiki/File:Hoyt_S_Vandenberg.jpg
    USAF, Public domain, via Wikimedia Commons
    Vandenberg

22. https://commons.wikimedia.org/wiki/File:Karl_W_Richter.jpg
    www.usafa.af.mil (archived version), Public domain, via Wikimedia Commons
    Karl Richter

23. https://commons.wikimedia.org/wiki/File:William_Pitsenbarger.jpg
    USAF Photo, Public domain, via Wikimedia Commons
    Pitsenbarger

24. https://commons.wikimedia.org/wiki/File:Duane_hackney.jpg
    http://afehri.maxwell.af.mil/images/afcross/hackney/hackney3.jpg
    Public domain   This image or file is a work of a U.S. Air Force Airman or employee, taken or made as part of that person's official duties. As a work of the U.S. federal government, the image or file is in the public domain in the United States.
    Duane Hackney

25. https://commons.wikimedia.org/wiki/File:Col_George_Day_official_portrait.jpg
    Bolling Air Force Base official site, Public domain, via Wikimedia Commons
    George "Bud" Day

26. http://www.af.mil/shared/media/photodb/photos/020925-O-9999G-015.jpg
    This image or file is a work of a U.S. Air Force Airman or employee, taken or made as part of that person's official duties. As a work of the U.S. federal government, the image or file is in the public domain in the United States.
    Robinson Risner

27. https://commons.wikimedia.org/wiki/File:James_H_Kasler.jpg
    USAF, Public domain, via Wikimedia Commons
    James Kasler

28. https://commons.wikimedia.org/wiki/File:Henry_Arnold_May_1911.jpg
    US Air Force Photo, Public domain, via Wikimedia Commons
    Henry H. "Hap" Arnold

29. https://commons.wikimecia.org/wiki/File:The_Wright_Brothers%27_First_Heavier-than-air_Flight_(7605918566).jpg
    NASA on The Commons, No restrictions, via Wikimedia Commons
    The First Flight, 1903

30. https://commons.wikimedia.org/wiki/File:Daniel_James,_Jr._in_front_of_his_McDonnell-Douglas_F-4C_Phantom.jpg
    The original uploader was Signaleer at English Wikipedia., Public domain, via Wikimedia Commons
    Daniel "Chappie" James

31. https://commons.wikimedia.org/wiki/File:Carl_Spaatz.jpg
    AFHRA, Public domain, via Wikimedia Commons
    Carl Spaatz

32. https://commons.wikimedia.org/wiki/File:James_H_Doolittle.jpg
    USAF, Public domain, via Wikimedia Commons
    Jimmy Doolittle

33. https://commons.wikimedia.org/wiki/File:Wolfenbarger_2012_2.jpg
    USAF, Public domain, via Wikimedia Commons
    Janet Wolfenbarger

34. https://commons.wikimedia.org/wiki/File:Jeannie_Leavitt.jpg
    USAF, Public domain, via Wikimedia Commons
    Jeannie Flynn Leavitt

35. https://commons.wikimedia.org/wiki/File:TSgt_John_Chapman_in_Afghanistan.jpg
    USAF, Public domain, via Wikimedia Commons

TSgt John Chapman in Afghanistan

36. https://commons.wikimedia.org/wiki/File:MQ-1_Predator_unmanned_aircraft.jpg
U.S. Air Force photo/Lt Col Leslie Pratt, Public domain, via Wikimedia Commons
The Predator Remotely Piloted Aircraft

JIM PFAFF

NOTE ON SOURCES

If we were walking through a museum, much of what I say as your guide would be the sort of knowledge that borders on folklore: "Things Airmen know." But some things I've written deserve sourcing, and you should know who deserves credit.

https://www.airandspaceforces.com/PDF/MagazineArchive/Documents/2009/August%202009/0809luke.pdf (Frank Luke)

Mets, David R. Master of Airpower, Presidio, 1988, (Spaatz's quotes on airpower's decisiveness and on attending so many Airmen's funerals.)

Sherry, Michael S. The Rise of American Air Power: The Creation of Armageddon. Yale, 1989 (p. 47-49, Rickenbacker and Mitchell quotes on carrying the mail.)

Bill Yenne, Hap Arnold: The General Who Invented the U.S. Air Force, p. 266 (Arnold legacies)

https://www.americanairmuseum.com/archive/person/michael-l-arooth

https://www.dm.af.mil/Media/Article-View/Article/314529/saluting-our-veterans-arizona-hero-arthur-j-benko/

Hinds, John W. Air Power History, Fall 2003 (Ben Warmer)

http://www.4thfightergroupassociation.org

https://cafriseabove.org/wp-content/uploads/2020/10/A-Short-History-of-the-Tuskegee-airmen.pdf

https://398th.org/History/KIA/index.html (WWII casualty figures.)

Copp, DeWitt S. Forged In Fire, Doubleday, 1982, p. 269 (Charles Kegelman)

https://www.303rdbg.com (Alan Magee, Joseph Sawicki)

Hoyt, Edwin P. The Airmen, McGraw-Hill, 1990, p. 44-47 (Colin P. Kelly)

Findagrave.com has information on many of those profiled. (Meyer Levin, Marty Sidener)

https://www.newspapers.com/article/fort-worth-star-telegram-john-landers-f/148862031/ (John Landers obituary, Fort Worth Star & Telegram,

September 15, 1989.)

Jones, Gregg, Most Honorable Son, Citadel, 2024 (Ben Kuroki)

Jeffrey Ethell, Alfred Price, Robert F. Dorr, Air Power History, Fall 1989. (Vogt on rescue.)

On the Lavelle Affair: https://apps.dtic.mil/sti/tr/pdf/AD1030383.pdf (The Lavelle Affair: An Air Force Case Study In Ethics by Kristina Ellis - School of Advanced Air and Space Studies Thesis, Air University, Maxwell AFB, AL, 2016

https://www.airandspaceforces.com/article/1106lavelle/

https://www.airandspaceforces.com/PDF/MagazineArchive/Documents/2007/February%202007/0207tapes.pdf

https://www.militarytimes.com/opinion/2023/09/13/why-i-serve/ (Gen Duke Richardson)

Robin Olds, Christina Olds, Ed Rasimus. Fighter Pilot. St Martin's, 2010. (On Olds' "middle finger" comment.)

"People In The Air Force" data from Air Force Magazine Almanac 1989 & 2025

Malachowski quote: https://www.aviationquotations.com/womenflyquotes.php (attributed to Seattle Post-Intelligencer, March 15, 2006)

https://www.airandspaceforces.com/article/silver-star-airpower-airmen-and-guardians-take-on-iran/ (Countering Iranian drone attack on Israel in 2024.)

Haulman, Daniel L. One Hundred Years of Flight: USAF Chronology of Significant Air and Space Events 1903–2002, Air University Press, 2003

McFarland, Stephen L. A Concise History of the Air Force. Air Force History and Museums Program, 1997

https://www.ibiblio.org/hyperwar/AAF/StatDigest/index.html.1945. This is the Army Air Forces Statistical Digest, my primary source for the "Numbers tell the story" section. You can find some of this information all over the internet, since at least 2011, usually with no attribution.

JET NOISE IS THE SOUND OF FREEDOM

www.ingramcontent.com/pod-product-compliance
Lightning Source LLC
Chambersburg PA
CBHW071151160426
43196CB00011B/2051